LOOKOUTS

Firewatchers of the Cascades and Olympics

Second Edition

Ira Spring and Byron Fish

THE
MOUNTAINEERS

Published by
The Mountaineers
1001 SW Klickitat Way
Seattle, WA 98134

0 9 8 7 6
5 4 3 2 1

Published simultaneously in Canada by Douglas & McIntyre, Ltd., 1615 Venables Street, Vancouver, B.C. V5L 2H1

Published simultaneously in Great Britain by Cordee, 3a DeMontfort Street, Leicester, England, LE1 7HD

Manufactured in the United States of America

Edited by Linda Robinson
Maps by Gray Mouse Graphics
Unless otherwise indicated, all photographs by Bob and Ira Spring
Cover design by Watson Graphics
Book design and layout by Gray Mouse Graphics

Cover photograph: *Fire lookout on Mount Pilchuck, restored 1990* (Photo by Ira Spring)

Library of Congress Cataloging-in-Publication Data
Spring, Ira.
    Lookouts : firewatchers of the Cascades and Olympics / Ira Spring and Byron Fish. — 2nd ed.
        p.   cm.
    Includes index.
    ISBN 0-89886-494-1 (alk. paper)
    1. Fire lookout stations—Washington (State)—History. 2. Fire lookouts—Washington (State)—History. 3. Forest reserves—Washington (State)—History. 4. United States. Forest Service—Officials and employees—History. I. Fish, Byron, 1909–1996. II. Title.
SD421.375.S67   1996
363.37—dc20   96–20760
                                                                                CIP

CONTENTS

Bench Mark Mountain Lookout

Anvil Rock Lookout

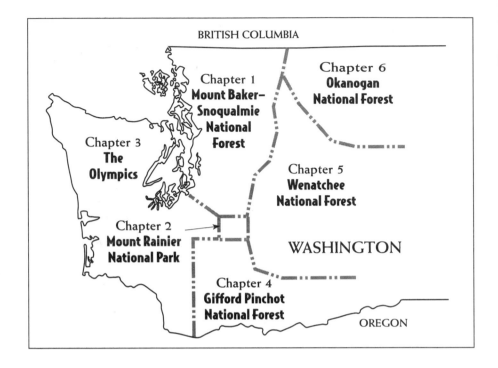

BRITISH COLUMBIA

Chapter 1
**Mount Baker–
Snoqualmie
National
Forest**

Chapter 6
**Okanogan
National Forest**

Chapter 3
**The
Olympics**

Chapter 5
**Wenatchee
National Forest**

Chapter 2
**Mount Rainier
National Park**

WASHINGTON

Chapter 4
**Gifford Pinchot
National Forest**

OREGON

3. The Olympics   91

Pete Miller's Treehouse

4. Gifford Pinchot National Forest   109

Mount St. Helens Lookout

5. Wenatchee National Forest 143

Domke Mountain Lookout

6. Okanogan National Forest 167

Monument 83 Lookout

PREFACE TO THE SECOND EDITION

When Byron Fish and I first began work on this book in 1980, the fire lookout system in western Washington was in its final decline. These observation posts, built and staffed by the U.S. Forest Service or the State of Washington, and which had served since the 1930s as an effective means of forest fire detection, had become all but obsolete. Mobile air surveillance and other new technologies, it was found, could keep far more efficient watch over the forest than a lone firewatcher in a tower atop a mountain. There was little reason to allocate federal or state funds to maintain their upkeep, so a great number of these historic buildings began to fall into disrepair.

Almost twenty years later, even fewer lookouts are in active use. However, while the federal and state governments still have no funding to maintain these buildings, a number of volunteer organizations have taken over their preservation. Many lookouts are still standing because of the efforts of volunteers, who do everything from reconstructing trails to moving abandoned buildings down from the mountains into museums. A huge debt is owed to these groups who keep alive this part of Washington history.

This new edition of *Lookouts* has been updated to reflect such changes, to report the status of these buildings a decade and a half after the book's inital publication. It also adds new anecdotes gathered from readers. Most of the letters we received after the publication of the first edition came from friends or relatives of the firewatchers mentioned in this book. Others were from present-day lookouts, hikers, or local historians who know the sites firsthand. All offered new or more detailed information about the history of the structures and the firewatchers who lived in them. Their data has been incorporated into the text and constitutes a major portion of the updated material in this new edition.

Several readers sent us information to update the book's original appendix of lookouts on the historic registry. Paul A. Swanson, for example, wrote that while geologically mapping the Goat Rocks Wilderness in 1982 he found what are probably the remains of the Lookout Mountain structure. In 1987 scoutmaster Niel Felgenhauer and his boy scout troop, hiking to Lake Caroline, missed a switchback and ended up in what they believe to be the remains of

Opposite: *Heybrook Lookout with Mount Index, left, and Mount Persis, right. Photographed in 1969. Trees are now covering the view.*

the Jack Ridge Lookout between Trout Lake and Jack Creek. We had originally listed the sites of these lookouts as unknown.

Two new appendixes have been added to this edition. The first gives driving and hiking directions to selected lookouts described herein. Many sites are accessible by car, others are at the end of an easy or moderately strenuous hike, but some should be attempted by serious hikers or climbers only. It must be emphasized that this is not a guidebook and these brief directions are given only as a starting point for those seeking out these historic sites. Many good outdoor guides (see Appendix 2) offer detailed descriptions of hikes to these areas and include such information as the precise level of difficulty and best times to visit. Know your skill level and what awaits you before heading out to any of these lookout sites.

The other new appendix describes volunteer organizations active in preserving lookouts. It is hoped that after visiting some of these spectacular sites, readers will be prompted to help the efforts of one of these groups.

Also new to this edition are icons (🏠) designating a lookout's inclusion on the National Register of Historic Places.

_____

It was after some urging from my sister Kay that I decided to collaborate with Byron—who was also my brother-in-law—to do this book back in 1980. Byron had a long-standing attraction to fire lookouts, as a freelancer having written stories about them as early as the 1930s for the *Saturday Evening Post*. I had no particular interest in these buildings, and I entered into the project with only mild enthusiasm. As it turned out, my time doing research for this project was the most fascinating summer I ever spent.

My initial encounter with lookouts had been in 1937. On my first day working at Paradise Inn, I ran up to 9,584-foot Anvil Rock. I talked to Wallace Meade, the man on duty, took a picture of the lookout with my Box Brownie, and ran back to the Inn in time to go to work. My next experience was in 1939 when my brother Bob and I climbed Mount St. Helens and Mount Adams. The building on top of St. Helens had fallen down and was covered with snow. The lookout on Adams was standing on the wind-swept summit and jam-packed with snow. I think one of us took a picture of the building, but fifty-seven years later cannot find it in my files.

In 1949, on the second portion of our honeymoon, my wife Pat and I accompanied The Mountaineers on their Glacier Peak summer outing. While camping at Image Lake, I did a *Seattle Times* story about a firewatcher on nearby Miners Ridge. A year or two later I did a story on Pat's uncle Mel and friends hunting deer on Mount Setting Sun. He and his party were packed in by horse, and had a deluxe camp just below Setting Sun Lookout. I got some great pictures from the top, but not of the building.

My only interest in lookouts had been for their trails which took me to dramatic viewpoints. I may have even braced myself against a building while taking a picture, but seldom took a picture of the building itself.

When we began work on this book in 1980, my job was to find or take photographs of the lookout sites or, if they existed, the buildings themselves. No list of lookouts was available, so I spent the spring of 1980 going over my 1930s maps looking for lookouts, and between Byron's and my own research, we located 466 sites.

That summer Pat and I visited fifty or more lookout sites. Some were on trails that had been abandoned so long they were overgrown and we could only find bits of tread here and there. On others, like Circle Peak, whose trail had been abandoned for 40 years, we found the tread in better shape than many maintained trails. I made an all-out effort to get pictures of the remaining lookouts and sites of others. The Forest Service was enthusiastic, and rangers did everything they could to help.

I found out that summer why these these odd little structures sitting high up in the mountains had held such fascination for Byron. Each had a story to tell—about a particular breed of outdoor enthusiast who thrived in the solitary, austere conditions of life in the mountains. There was "Lightbulb," the friendly young man we met in 1980 who manned Green Mountain in the summers and spent his winters in India studying eastern religions. My wife's brother Ed and cousin Barney Douglass told us of their summers on Desolation Peak and Devils Dome in the 1940s where they hiked 19 or 20 miles through magnificent forests to get to their posts. One of Byron's favorite stories was about Nels Bruseth, the lookout on Pugh Mountain who ran down the 6-mile trail every Saturday evening to take his girlfriend to the weekly dance in Darrington. After the dance he would climb up the 6,000 feet to be back on his job at daybreak.

In truth, not *all* firewatchers flourished in their mountain towers. One of Byron's *Saturday Evening Post* articles tells of a less than happy young man who kept watch at the Three Fingers Lookout in the 1930s. The structure on the 6,854-foot peak is surrounded on three sides by 1,000-foot drop-offs and on the fourth by the Queest-Aib Glacier. A Forest Service rescue party had to be dispatched to Three Fingers one foggy day because the firewatcher had developed such a bad case of acrophobia he couldn't get off the cabin floor.

All these stories, with their spectacular mountain settings and cast of heroic, eccentric, and ascetic characters, are what Byron loved about lookouts.

In March 1996, before work on this new edition was completed, Byron died. An accomplished writer and journalist with a wonderful tongue-in-cheek sense of humor, Byron deeply loved the forests and mountains of the Pacific Northwest and spent many hours hiking and camping in their confines. He was captivated not just by the region's beauty but also by its history, and in numerous writings over the years he sought to make that history come alive for his readers. I think he would be pleased that our book on the firewatchers and their lookouts, in print now for fifteen years, will in this updated form reach an even greater audience.

—Ira Spring

Aerial view of Mount Constitution on Orcas Island, a Department of Natural Resources lookout built in 1936 by the CCC. Mount Baker is in the distance.

PREFACE

B ack in the 1930s the still-growing lookout system seemed like a stable
part of the fire protection program of the forests of Washington. Things
like helicopter surveillance, transistorized radios, smokejumpers, fire
retardant bombs, infrared heat sensors, and even a road system penetrat-
ing all but the most remote wilderness areas were far in the future. None of us
expected the future to arrive so soon.

With few exceptions, the lookouts featured herein are those built and
manned by the United States Forest Service in its five national forests in
Washington. In planning this book, our original estimate was that some 200
lookouts had been used in the state. Further investigation revealed records of
more than twice that number, and we're still counting.

The state also built its own network of observation posts, scanning state
and private timbered property no less important than that of federal lands.
Close to 100 such lookouts were built, most of them administered by the
Department of Natural Resources. Only a handful are still used, mostly in
eastern Washington.

It soon became obvious that there would not be room to give complete
information about all or even most of the lookouts that had existed in
Washington's Cascades and on the Olympic Peninsula. We decided we
would have to single out those that had the most interesting history or
anecdotes behind them, or whose sites were scenically spectacular, for de-
scriptive treatment.

This proved to be a reasonable decision, in fact the only one possible
short of a major research project that would end up with more details than
anyone would care to read, about an era for the most part now in the past.

Considerable interest was shown in the project from the start. Files of
the U.S. Forest Service, the National Park Service, and the state Department of Natural Resources were willingly opened to us. If these files did not
turn up what we were looking for, the agencies often could steer us to some
old-timer who might have the answers. We hope readers will continue to
volunteer information or corrections from firsthand knowledge. We also
welcome dissenting opinions—the observations and anecdotes herein came
from individual recollections which may differ from those of others recall-
ing the same places and times. Such material will constitute another source
of information for the regional historians of the future, and the publisher
will be glad to receive it.

None of these lookout locations, state or federal, was chosen for ease of access or safety. Due to their function as watch stations, most have impressive views, but some are downright dangerous to reach, and visitors must be cautious. Many of the old lookout buildings have been demolished or burned by the Forest Service, and few of the ones that still stand are free of the effects of disuse, severe climate, and vandalism.

Then too, lookout sites are often located where the land has limited abilities to absorb human impact. These settings dramatize the determination of those who faced the rigors of their work during the era of the lookouts. In a few cases, the structures and materials that remain provide insights of historical or archeological value into the individuals and techniques of the past. It is important for visitors to realize that what is left in and around old lookouts tells a valuable story, one that can be lost forever if materials are moved, altered, or removed as "souvenirs."

That is one reason we have refrained from making this work a guidebook. However, we have added an Historical Registry of Western Washington Lookouts in the back of the book, giving information as complete as current records make possible on the lookouts built and used by the Forest Service and the Department of Natural Resources. Location data and access status are included, as well as building dates and miscellaneous notes on every lookout site we could find. It is hoped that those whose interests are sincere will thus not be completely dissuaded from tracking down these historic sites on their own.

Thanks to all who generously gave their time to interviews or to digging up old photographs or records. Special acknowledgment goes to Keith McCoy of Trout Lake, a third-generation resident of that area whose hobby has been its history; and to F. E. "Ernie" Childs, who was in the woods when we tried to track him down, but who cheerfully allowed Keith to pump him for details and forward his contributions.

An ardent outdoorsman who makes a point of learning all he can about the places where he hikes, Dr. Fred T. Darvill, Jr., of Mount Vernon was more than willing to share his knowledge with us. Darvill has written about some North Cascades trails and lookouts himself, and one of his books, *Mountaineering Medicine* (published by the Skagit Mountain Rescue Unit), has become a standard in its subject.

Most of the others who helpfully reminisced are named in the text. We hope that what they said doesn't get them into too many arguments with friends who remember the situation or the incidents a bit differently.

Note: Forest Service personnel did not live or work "in" a district or national forest, any more than a seaman on a ship would say he "went down to the basement." The word was "on." Other people, of course, are simply "in" whatever area they reach. As a compromise between that popular usage and the professional version, "on" is generally used in this book whenever a Forest Service veteran is quoted, directly or indirectly.

INTRODUCTION

There were so many trees all across America when the United States was formed that it wasn't until a century later that anyone had reason to worry about them. Forests were a bountiful and important natural resource, and a seemingly endless one. Only when their loss became threatened did anyone look upon their future as a matter for public policy. The American Forestry Association was formed for this purpose in 1875 and could point to a recent but terrible example of how fast a forest could be lost. The Peshtigo fire in Wisconsin in 1871 had swept across 1,300,000 acres, and in the process taken 1,500 lives.

In 1891, the U.S. Congress gave the president authority to withdraw public lands and establish forest reserves. Benjamin Harrison started off with 13 million acres adjacent to Yellowstone, which had become the country's first national park in 1872. Grover Cleveland followed in 1893 with 21 million acres here and there, including the vast Pacific Forest Reserve over much of Washington State.

That's all they were at first—"forest reserves." The land was put on hold

Bearhead Mountain Lookout and Mount Rainier (Photo by R.N. McCullough)

under the General Land Office, Department of the Interior. There was no plan for fire protection or any program for timber harvest. The main accomplishment was that in the reserves, timbermen were stopped from pulling one of their old tricks. "Settlers" had been hired by timber interests to take up the full homestead allotment in the richest forest land; as soon as they gained title, they sold it for a few dollars, by prearrangement, to their employers.

The Europeans had had to protect and manage their own forests for a long time. Gifford Pinchot, a Yale graduate, attended schools of forestry in France, Germany, Switzerland, and Austria; when he came home, he had conservation ideas that were extremist in the United States. He got to practice them for the third-generation multimillionaire Cornelius Vanderbilt, who had a private reserve on his vast estate at Biltmore, North Carolina. Later, Pinchot founded the School of Forestry at his alma mater, Yale. In 1896, he became a member of the government commission planning how to handle the reserves, and in 1898, at the age of thirty-three, he was appointed head of the Division (later Bureau) of Forestry.

Politically, his arrival on the scene was timed just right. In 1901 along came Teddy Roosevelt, who would rather have been outdoors than president. Since he had fallen heir to the big job, however, he added 132 million acres to the Forest Reserve System. With Roosevelt and Pinchot sharing a mutual enthusiasm over what should be done, Congress transferred the reserves from the General Land Office to the Department of Agriculture in 1905. The Bureau of Forestry became the U.S. Forest Service in 1907, with Pinchot still as the chief—a position he held until 1910, when he moved on to teaching and conservation interests.

Shortly thereafter, the first fire lookouts were initiated. It began with the natural thing to do: climb to a high point for a better look around. Summer camps with a tent ("rag houses") and sighting instruments were established, followed by cabins in the more strategic spots. About the time of the First World War, the assault on the peaks began in earnest. Heroic labors were needed to haul materials to the top of a mountain to build a lookout cabin and to string telephone wire up to it before radio became common.

In national forests all over the West, the building boom reached its climax between 1929 and 1935, when ranger districts were completing lookouts at the rate of two to four a year. They were placed, wherever possible, so at least two stations could overlap surveillance of the same territory and thus accurately pinpoint the location of a fire.

A second President Roosevelt gave a boost to the program. The Forest Service already was doing its best in 1932 to put unemployed men to work, which may have given the New Deal administration its idea. Probably, there was also enthusiastic advice contributed by the then Governor of Pennsylvania, Gifford Pinchot.

The new president and his Secretary of Agriculture, Henry Wallace, asked Robert Y. Stuart, chief of the Forest Service, if he could use an extra 25,000

Winch used to haul supplies to the lookout site on Pugh Mountain

young men on the payroll. Stuart, happy to accept, was somewhat taken aback when the administration amended the number to 250,000.

The CCC (Civilian Conservation Corps) was in business from 1933 to 1942, with the peak enrollment of 500,000 in 1,500 camps across the nation coming between 1935 and 1937. In its nine years, the CCC employed two-and-a-half million people, who worked on both Forest Service and National Park Service land. They built 60,000 miles of trail and 600 lookouts.

The trails and the lookouts, whether built by the CCC or the Forest Service, were to be a permanent part of the fire protection system. By the Second World War, the building period was over for the most part, except for the modernization of old stations or the building of a few new ones made necessary by such developments as timber sales or big burns. Sales in otherwise untouched valleys sometimes meant erecting a lookout for close scrutiny of a logging operation for the fires it could cause. An after-the-fire lookout was to guard against a reburn in a recently devastated area that might be reseeded from adjoining timber stands.

The war at first delayed whatever changes might have come about under a peaceful course of events. The uneasiness about the perceived threat from across the Pacific led to the year-round manning of some lookout stations as part of the military's Aircraft Warning System. In some cases, husband-and-wife teams were hired for these stations. And in general, although a few women had been lookouts each year almost from the beginning, the Forest Service began hiring more women lookouts for its normal summer work, since so many young men were in other kinds of uniform.

Once the war was over, leaving its legacy of technological advances, change speeded up. Airplanes and helicopters were considerably different vehicles from what they had been in the 1930s, and parachuting supplies and men was no longer in an experimental stage.

A genuinely portable two-way radio, an instrument the Forest Service had been developing before the war, had been seized upon by the military and, with the unlimited money of wartime emergencies, rapidly improved upon for battle use. Afterwards, when the Forest Service could have those radios as government surplus, it was no longer necessary to keep repairing telephone lines from ranger stations to lookouts.

Throughout all this, lookout cabins themselves went through several architectural periods. The first ones were free-style, anything that could be put together to serve the dual purpose of a place to live and an observation point. That combination produced some odd-looking structures, such as a three-story tower with storage on the bottom floor, living quarters in the middle, and a work room on top. None of these earliest posts seems to have survived, except in photographs.

Standardized cabins appeared in California before they were adopted in the Pacific Northwest. The first ones were called D-5s, being from California's District 5. They were 12-by-12 feet, with an observation cupola on top. In general, a cupola pegs a building to the 1920s, although a few were built after that decade.

Next came a 14-by-14-foot cabin with a gable roof, so simple in design that with its shutters closed it looked like a child's drawing of a house. They were not very popular and, at least in the Darrington Ranger District, were derisively nicknamed "grange halls."

The pyramid-shaped "hip roof" building arrived about the same time, in the early 1930s, and continued to be built into the 1940s. It was designed for the snowy Northwest Region 6, so it was called the R-6 model (or sometimes the D-6, for District 6). Last to arrive on the scene was the flat-top, introduced in the late 1940s. The result of more sophisticated building techniques and materials, it allowed the designer to overcome the peril of heavy snow on the roof. No great number of them were built, because by then lookouts were starting to be phased out.

Although the standard size of the non-cupola cabins was 14-by-14 feet, there were exceptions forced by the space available—12-by-12 or even 10-by-10. Domke Mountain's tower in the Wenatchee National Forest, around 100 feet high, was topped by a 6-by-6-foot cabin with just room enough for the firefinder, a small table, and a chair. The lookout had to descend to ground level to eat or sleep.

The most important piece of equipment a lookout had was the Osborne firefinder, the device used to get precise bearings on any fire or smoke spotted so the location could be relayed to the district ranger station. For the most part the firefinders were manufactured in Portland, Oregon. A movable device not unlike a rifle sight could be turned by the lookout to take a bead on a fire, while

Four styles of lookout architecture: Top left, *A D-5 with observation cupola at Jolly Mountain.* Top right, *A "grange hall" on Copper Mountain.* Bottom left, *A "hip-roofed" D-6 on Huckleberry Mountain.* Bottom right, *A flat-top at Bunker Hill* (U.S. Forest Service photos)

an immobile ring around the outside gave the azimuth readings of the location. If more than one lookout station could get a reading on the same fire, the district ranger station could triangulate the exact location.

In a typical layout, the firefinder took up a 2-by-2-foot space in the center of the cabin, with the telephone or two-way radio mounted beside it. Bed, table, and stove filled three corners, with the door on the fourth. Cupboards below the windows held supplies. The building was protected by lightning rods grounded to all sides, and the telephone also had a grounding switch. On peaks subject to numerous lightning strikes, the lookout could help protect himself by standing on a low stool whose short legs were insulators.

Wherever it was reasonably available, wood was used in the airtight stove

Aircraft Warning System observers at Rinker Point Lookout near Darrington, January 1943 (U.S. Forest Service photo)

for both heating and cooking, and each lookout was supposed to leave a cord split for his successor. On high rocky peaks far above timberline, the stove ran on kerosene. For additional warmth, the lookout dressed in woolen shirts and longjohns.

A young man who hoped for a career in forestry might begin on a trail crew, advance to manning a guard station, and then reach his first starring role as a lookout. Although he was not in fact very far along in the ranks of Forest Service personnel, he received a little more pay and, more importantly, a certain recognition. He had a peak of his own for the summer, and what with all the reporting back and forth, everybody in the district knew him as an individual up there.

To the public, "lookout" and "hermit" were synonymous, even though those on the most accessible stations sometimes felt they were running a continuous open house. Few were so remote they had no company at all, because hermits have always fascinated people, and curiosity led hikers to mountain tops just to find out what sort of person lived up there all by himself. Thus, lookouts were resigned to one inevitable question: "Don't you get lonesome?"

Visitors were generally surprised to hear that there was no spare time for loneliness. Even on zero-visibility days, there was water to fetch, wood to split, meals to prepare, dishes and clothes to wash, and four walls of windows to clean (inside and out) at least once a week. When it was raining, conditions were considered excellent for doing trail maintenance work.

James Currie, who began at age sixteen and spent six summers in the Wenatchee National Forest as a lookout, wrote in a reminiscent letter, "There was never a feeling of loneliness or solitude, but rather one of being home. It was the same each year, the 'going home' feeling was much stronger when moving onto the station than it was on coming down in the fall."

Nevertheless, not everyone who tried to be a lookout could stand the job. The biggest turnover was on peaks subject to frequent and violent lightning storms, during which a cabin might be struck a number of times. It was at the heart of a lookout's job to stay on duty during these terrifying storms, watching for lightning strikes that might turn into fires.

No amount of forewarning could prepare beginner lookouts for the actual "combat" experience. When it happened, some were convinced they had survived by luck alone, and they asked to be relieved. At least one fellow—on Remmel Mountain in the Okanogan National Forest—didn't wait for formalities; he just took off running down the mountain. But no one we talked to knew of a lookout who was killed by lightning, though one man said he had heard of a Canadian lookout and his wife who were crippled because of an improperly grounded cabin.

Usually, though, attrition among lookouts was simply one of personality. Most rangers remember hiring an extrovert who worked happily and efficiently on a trail crew, joking with everybody. Then he would decide he wanted to move up to lookout and was given the chance. A week of seeing no one else brought on cabin fever and a request to switch back to his old, more sociable job.

In talking to those who had served as lookouts, it was quite evident that having the necessary temperament was in no way allied to an antisocial attitude. On the contrary, the best lookouts were those so at ease among their fellow beings that they were equally at ease by themselves.

Some of the sources we interviewed turned out to be relatively young men and women who had been lookouts during their teens and early twenties and then gone on into careers not related to the outdoors. To this day they fondly recall all the events of a summer, along with the names of rangers, packers, and their fellow lookouts on other peaks. It was obvious that the time they had spent in self-reliant living, whether for a season or during many years, was an experience they will cherish for the rest of their lives.

Novelist Jack Kerouac manned Desolation Peak and poet Gary Snyder spent a summer on Sourdough Mountain. A New York opera singer, Dan Ames, entertained wildlife on Burley Mountain while practicing for a new role. On the 20-foot lookout tower of Bonaparte Mountain, a music student was building a grand piano from the parts friends brought up the 4.5-mile trail piece by piece. The man was not certain how he would get the piano out the door and down the trail when it was finished.

There are few lookout stations manned today. In fact, few of the buildings are left at all, because those abandoned were torn down or burned by the

Green Mountain Lookout. White Chuck Mountain is in the distance.

Forest Service. Besides aerial surveillance and other technological advances, economics had entered the picture by the 1960s.

In the 1930s a ground cabin could be built for $500 and a tower for $1,000 or less. Maintenance, packing, and the lookout's wages might come to $1,000 for a summer, out of which the lookout was lucky to draw half if he also put in early season work on trails before he moved to the peak. A typical wage was $100 a month, less $5 rent for the lookout cabin. (No matter that his quarters were put there for business reasons, it was government policy that anyone occupying a federally owned building must pay rent.)

In 1950 the 12-by-12-foot cabin with cupola on Kelly Butte, Snoqualmie

National Forest, cost $3,461. When the old Granite Mountain lookout above Snoqualmie Pass was replaced by a 14-by-14-foot cabin in 1956, the investment value was entered on the books as $7,608. In 1964, a 14-by-14 cabin on a 70-foot tower at Heybrook came to $22,028.

Prices of everything else had gone up, of course, in thirty years, but not by twenty times. The Forest Service was in a losing battle against something more intangible than the rising cost of materials and labor. There was no longer a governmental project like the CCC to help out, and outside contracts that once were accepted as a challenge in themselves had become just another bid on a government job.

Les Larsen of Olympia, who retired from the Forest Service in 1975, was assigned to do a survey on investment-versus-return of lookout stations in the Olympic National Forest during 1965 and 1966. By then it cost an average of $8,000 annually to keep each lookout open. He studied a sixteen-year record of initial fire reports and found that some lookouts had made none, at least in recent years. That influenced a decision to phase out the Olympic stations from the remaining dozen in 1966 to none in 1975. By that year, Larsen estimated, it would cost $12,000 to build a lookout and as much or more to man, supply, and maintain it for a season.

Of course, it wasn't just that modern aerial surveillance could substitute for lookouts. Unlike in the old days, the woods were now full of people— loggers, hikers, and fishermen— all traveling on a network of roads and trails that did not exist in the past. They all knew a fire when they saw it and knew where to report it.

Another thing had changed, Larsen added somewhat gloomily—the spirit behind the whole system. Under the more primitive conditions of the past, a lookout had to be a dedicated person, willing to undergo the physical exertion that went with the job, self-sufficient for weeks at a time if necessary, and on duty at any or all hours.

"In the end," he said, "we had college students who appreciated the solitude but were too busy writing a thesis to notice fires. And drive-in stations where the lookout could serve a shift and go home. Or sneak away to town for a beer."

Vandalism had always been a problem and reason enough to remove seldom-used lookouts. In 1992 Red Top in the Wenatchee National Forest was viciously and destructively vandalized. The vandals tore off shutters, broke railings, destroyed the furniture—including the propane stove and refrigerator—and stole the firefinder.

Four enthusiastic volunteers from Kittitas Search and Rescue—Don Kerns, Art Pieters, Warren Burger, and Bob Jackson—are working to restore the building as a visitor center.

The Forest Service might have left many abandoned stations on standby for emergency use if it had not been for an act passed by Congress in 1965. It opened the way for citizens to sue a federal agency for injuries suffered on government property—presumably through no fault of the victim. However, with that sort of law on the books, it was not likely that everyone who fell through the rotted floor of a long-deserted building would exclaim, "Oh, pshaw, I should have watched my step!" and let it go at that.

Considering the potential for falls from lookout towers or peaks with buildings that might be deemed in court an "attractive nuisance," Forest Service headquarters decided that it was not going to be sued for that kind of accident. The word went out: if a lookout station has no further use, tear it down or burn it—just get rid of it.

One by one they were removed. Many a ranger, packer, or lookout who had helped build a station at the risk of his neck presided over its demolition years later—often at the same risk.

Will manned lookouts become entirely a thing of the past? Not likely, was the word in a 1974 Forest Service study. According to that study, in 1953 there were 5,060 "permanent" structures in national forests across the country. From then on their number steadily declined, but those that remained consistently spotted more fires than did aircraft. That was 1974. Thirty years later "not likely" has become "likely," as of the forty-seven buildings still standing, twenty-four are still maintained and manned by government employees. The others are either abandoned or maintained and manned by volunteers.

"The reduction in lookouts has had little effect on their value in the fire detection system," the 1974 report read, "possibly because those eliminated were poorly located to begin with." Or, it could be added, because the circumstances that led to the placing of a lookout had been changed by logging, permanent roads, and other factors that made its original function no longer necessary.

The surviving lookouts were also reported to have kept an edge over aircraft in several other ways. They acted as weather stations, recorded the path of lightning strikes, and provided human contact with forest users.

"There is nothing to indicate that they will be eliminated or replaced with unmanned, fully remote systems," the study concluded. "Such devices are not available and it does not appear they will be available in the foreseeable future to replace present manned, fixed detectors."

Yet considering the sophistication of exploratory spacecraft, orbiting weather stations, and the increased use of mechanized labor, it is possible that someday lookouts will be marvels of circuitry, standing 24-hour watches and scanning the countryside with built-in cameras and infrared heat detectors.

It is also possible that when technology reaches that point, it will be less expensive to hire humans for a seasonal job.

And since packstrings run on native renewable resources instead of petroleum-based fuel, maybe they too will make a comeback. So we cannot limit speculation to the assumption that the only direction the future can take is into science-fiction gadgetry. In the years ahead, hikers may see history coming as well as going.

1

MOUNT BAKER–SNOQUALMIE NATIONAL FOREST

It was great. In the 1930s and '40s one could have all the advantages of living in a metropolitan area and still start to backpack or mountain climb in less time than it took to attend a symphony concert. Mount Baker and Snoqualmie National Forests covered nearly 5,000 square miles of some of the most rugged mountains in the then forty-eight states, beginning only an hour's drive away from the populous Puget Sound country.

When you couldn't manage a week for exploring, you still could take off after work or classes on Friday, camp at a trailhead, and spend 2 days hiking

Barney Douglass, the lookout at Devils Dome. The photo was taken about 1938 by a government packer.

A "rag house" on Pugh Mountain, about 1918 (U.S. Forest Service photo)

among the peaks without seeing another person. You might cross paths with
a Forest Service trail crew, but they were not included in the human count.
They were a form of wildlife, even more interesting than bears. They be-
longed in the evergreen jungle.

A nineteen-year-old on his first trail job was still a Forest Service Man and
consequently one who rated respect. It was taken for granted that any Forest
Service employee was an expert axman, never got lost in the woods, could start
a fire with two sticks, and could out-pack and out-hike just about anyone else.
His very employment by the Service made him a professional Woodsman.

If you met up with a lookout, either on the trail or in his lair, never mind
that there had been 2 days of clouds and rain; the trip was a social and educational success. He could name all the peaks in sight (if any were), and he
often turned out to be a student from one college of forestry or another, full
of general information.

The ultimate excitement was to come upon the district ranger himself.
God made the mountains and the ranger was his Moses, down from the peaks
with the Commandments for All Outdoors. Or at least all the outdoors for
several hundred square miles around. A ranger who ascended from his district station to the high trails and lookouts was on business, and he looked
like it. He dressed, moved, and spoke with an air of authority that left one a
bit awed.

National management of forest resources was rightly the province of
trained administrators and scientists, but the foundation of the system de-
pended upon the residents of each region. They were a sparse population,
living out their lives in the valley hamlets and farms, roaming the mountains

to hunt, or just exploring out of curiosity. They knew the country around as well as they did their own backyards.

So the ranger often was accompanied by an older man, muscled with steel cable, who tended to speak softly and in few words. He was most likely born "on the district," and even back then was the second generation to work for the Forest Service.

The Darrington District was not unusual. Nels Bruseth was born down Stanwood way and never lost his Norwegian accent, but he lived in Darrington and worked for the Forest Service from his teens on. His knowledge of forestry came from experience. Like his counterparts on many another district, Bruseth never became a ranger, but he served as Number Two man for a succession of them. He was well educated from studies of his own, having a wide field of interests, and he compiled historical data on the region.

Bench Mark Mountain Lookout (U.S. Forest Service photo)

Bruseth wasn't big, but the wiry outdoorsman became a legend when talk turned to hiking feats. He gained his reputation early, as a lookout on Pugh Mountain. In courting the girl who was to become his wife, he took off at day's end and hiked 6 miles to the bottom of the mountain and another 8 to her home. Reversing the route, he was on the job at sunrise.

Another lookout who ran the Pugh Mountain Marathon was Harland Eastwood, but he did it for sport. On a cloudy afternoon with nothing to see, he trotted down to play a baseball game in Darrington. He looked upon the preliminary 14 miles as a nice warm-up.

The Bedals, of part Sauk Indian ancestry, were born in the area, and their name is pinned on a peak, a creek, and a campground on the upper Sauk River. Harry was another iron man, and two of his sisters, Edith and Jean Bedal Fish, were packers for the Forest Service. Jean also was a lookout.

One dedicated veteran of the Service traveled on four feet. Clinton ("Dutch") Tollenaar, a former packer in the Darrington District, had a dapple-gray mule named Mabel. He tried to retire her when she was thirty years old; but after the packtrain moved out to supply a lookout, Mabel wriggled under a fence, caught up, and took her rightful place at the head of the string.

Walter Anderson calling the district ranger from Noble Knob Lookout (Photo by R.N. McCullough)

There she balked until she was given a pack.

When Norm McCausland was a teenager, he did not settle for weekend hikes in the mountains. He went into the Skykomish region in 1925 and stayed there until 1969, working for the Forest Service. In retirement he still lived in Skykomish, another example of an area resident who gave the Service continuity even when rangers moved from one district to another or into higher office.

As protective assistant and eventually fire control officer, McCausland was a frontlines boss for many years. In the 1941 Rapid River fire, personnel were brought in from other districts and Wenatchee National Forest, but the main force of 1,500 men had to be hired from urban areas, large and small. McCausland and Nevin McCullough, the White River district ranger who had rushed in reinforcements, both recalled a labor dispute that raged along with the fire.

An "organized" crew was one that supposedly had experience in the woods, most likely as loggers. They were hired as a package complete with leader and they were paid thirty-five cents an hour plus grub. Individual recruits got thirty cents and food. With bigger-city sophistication, crews from the west side of the Cascades always joined up as an organized group. After a few days on the fire, eastsiders discovered they were being shortchanged a nickel an hour and threatened to quit.

His budget already up in flames, Herb Plumb, the Skykomish district ranger, was dismayed by the outrageous wage demand. "What'll I do?" he pleaded. The two Macs suggested that he either try to recruit 500 or more instant replacements or pay the difference. He groaned and paid. One ideal, the fight to save our national forests, lost out to another: equal pay for equal work.

Man-made fires were not always caused by carelessness. Wherever sheep or cattle roamed, a fire of unknown origin could result in additional grazing land. Then too, throughout the depression of the 1930s, fires required hiring of help to suppress them—when even thirty cents an hour was better than nothing. Mysterious fires broke out on all districts.

According to a story picked up from a couple of old-timers, one fire control officer studied the fire pattern and hired a local resident, otherwise unemployed, to watch out for arsonists. The incidence of fires was greatly reduced thereafter.

What else the guardianship of the forests meant, aside from building trails and watching for fires, was revealed by the telephone—as a former smokechaser in the Darrington District recalled from listening into conversations between lookouts, trail crews, and the ranger station.

During a dangerously dry weekend the ranger station asked a lookout, "Who's up the trail today?"

"A young couple who stayed at Boulder Ford lean-to last night," one lookout replied. "Heard them listening in on the phone. They must be at Goat Flat by now. I also spotted three high school kids but I recognized one as a local boy. They're just hiking."

A second voice came in. "A couple of fishermen went to Boulder Ford this morning. They're from Everett, according to the fire permit they got here."

The man at the ranger station cut in. "The packer came down last night. He said he found a camp 3 miles up the river. Whose camp is that? Did they get a fire permit?"

"No, never saw them, but the car's parked at road's end. Back seat's full of gear so I'd say there are two men. One guy works for a coffee company; the brand name is on the door. I took the license number."

"Better go bring them in."

Through his glasses, another lookout spotted a party of four trudging up the valley below. He told a trail crew farther on that company would be arriving. The hikers had not seen a soul in half a day, so the faster pair were astonished when they were greeted with, "What happened to the other two? Did you lose the guy in the green sweater?"

There were eleven persons on 17 miles of seemingly uninhabited trail that day, each group thinking it was lost to the world. In case of trouble, though, the Forest Service knew where they were.

All the above is in the past tense, as though everything is different today. It is true, of course, that another half-century of human activity has wrought changes. The once-graveled roads are now blacktopped highways, and hundreds of miles of gravel and dirt roads branch off from them, up mountainsides and river valleys that once had to be hiked. The packhorses are gone and the trail crews use chain saws. You don't often get to talk to a lookout because most of the people doing that work fly overhead in a light plane or a helicopter.

Still, the setting is the same. The Cascades stand solidly in the same place. Mount Baker and Snoqualmie National Forests were merged partly because of Baker's reduced territory, but the land removed went into the new North Cascades National Park. Valleys and mountainsides have been denuded by loggers, but large areas have been set aside under wilderness protection.

Solitude may not be found quite so readily as in the past, but it is not a lost cause. There are trails you can have all to yourself on weekdays, especially if they don't lead to fishing waters. A number of those mentioned herein are not much traveled because they led to lookouts that no longer exist. There are only traces to follow, and traces to find.

Church Mountain

Elevation 6,100 feet

Church Mountain, near the western boundary of the national forest between the Nooksack River and the Canadian border, has several summits, the highest of which is 6,315 feet. The lookout, a cupola-type put up in 1928, was on the rocky top of a 6,100-foot peak. In addition, there was a trail and a telephone line to West Church Mountain, elevation 5,610 feet, which provided the lookout with another vantage point if he needed it.

What he saw was the Nooksack Valley stretched out east to west, the town of Glacier almost a vertical mile below, and a ring of mountains—the

Aerial view of Church Mountain showing the storage shed at lower right and the trail zigzagging past the outhouse to the lookout building on top (U.S. Forest Service photo)

Sisters, Mount Baker, Shuksan, Icy, Ruth, and the Border Peaks. Down the north side were the two Kidney Lakes, buried under snow until August.

Keeping a building on Church Mountain was a problem. Heavy snow loaded it down and ice built up on the supporting cables until it almost collapsed the structure. When the cables were loosened, the building shifted on its foundation.

Actually, the same problem existed at many a cabin that had to be guyed in place. The laws of physics can be a nuisance. Wood (as in a building) expands in damp weather and shrinks when it dries. Steel (as in a cable) contracts in cold temperature and expands when it is hot.

Tensions had to be adjusted for the season, and it was not a job to be entrusted to a novice. With one material expanding and the other contracting under conditions that could range from sub-zero to plus-100-degree temperatures, setting the correct tension was somewhat akin to tuning a violin.

The lookout on Church Mountain was removed in 1967, and thirty years later a half-collapsed storage shed and an outhouse are all that remain. The site is fairly clean, but rusting cables and hunks of metal still dangle from where they were shoved over the cliffside. The 4 steep miles of trail climb the south side of the mountain.

Left, *Storage shed at the end of the horse trail. The roof was sloped to the ground so avalanches would slide off. Marmots live under the shed now.* Right, *Narrow-gauge outhouse. These two buildings can be seen in the aerial photograph,* opposite.

Winchester Mountain 🏚

Elevation 6,521 feet

O ver and over, the descriptions of trails in this book say they used to be longer or that they have been erased by logging roads or abandon-ment. A definite exception to that kind of statement is most of the climb to Winchester Mountain. Few routes have been so long established on the same path. Even when bulldozers finally arrived in 1949, they simply pushed their way up 6 miles of switchbacks, following a packtrail that had been there for nearly half a century.

The packtrail was built by miners who, between 1902 and 1906, carried in tons of supplies and machinery for a stamping mill at Lone Jack Mine, and brought out at least $300,000 worth of gold. Beyond Twin Lakes, in a saddle between Winchester and Goat Mountains, they went through Skagway Pass and curved south around the side of 6,891-foot Goat Mountain to the mine.

The 6-mile climb to the lakes was just the steepest part. They already had trudged 21 miles from Maple Falls on the Nooksack River. Snow blocked the upper levels eight months of the year. Under such conditions, Lone Jack was "temporarily" closed, to await better accessibility. By the time a "mine-to-market" road was built, the mine and mill long since had fallen into ruins.

For the next fifteen or twenty years the road allowed motorists to drive to Twin Lakes, inching forward and backing up in two or three maneuvers to get around switchback corners. It also brought the Winchester Mountain Lookout within 2 miles of supplies by truck.

The road to the lakes is maintained mostly by a few miners who go to Lone Jack to meditate about today's price of gold. It can't be counted on to give passage to ordinary cars, but even so Twin Lakes and Winchester are popular destinations for hikers.

A 1990 photograph of abandoned Winchester Lookout. Since then volunteers have restored the building.

Trail to Winchester Lookout, with Mount Shuksan in the distance

The lakes are in a park-like, subalpine setting, with masses of heather and flowers. Their idyllic appearance is enhanced when seen from the top of Winchester, 1,300 feet above. The mountain is about 3 miles from the Canadian boundary, which is marked by the Border Peaks—Canadian Red, Border Red, and American Red. In the 360 degree panorama, Mount Baker and Mount Shuksan appear in full glory.

Small tarn on Easy Ridge, with Mount Shuksan in the distance

Easy Ridge and Copper Mountain

Elevations 5,640 and 6,260 feet

Easy? The lookout and the ridge it stood on must have been named ironically. It was never easily reached and still isn't, now that Easy Ridge is in the North Cascades National Park and safe from more logging roads. To get there, one must depend on feet for 14 miles and climb to more than 6,000 feet.

Built on the 5,640-foot shoulder of 5-mile-long Easy Ridge, the lookout stared into the headwaters of the Chilliwack River. It was just a 1,000-foot climb to the 6,613-foot ridge top and a view into the Baker River watershed. The lookout was constructed in the mid-1930s for emergency use but was never manned. A ranger who visited it in 1969 said there were still cedar shingle shavings in a kindling box, left by the workmen who built it.

In 1934 Wesley Brown helped string the telephone line, clear the trail, and shingle the roof. He believes the carpenter was Paul van Cruyninger.

The lookout on Easy Ridge was put there because it had a better view of the Chilliwack Valley than did Copper Mountain to the north. The two lookouts were 15 trail miles apart but less than 2 miles by air—almost within shouting distance of each other.

Copper Mountain Lookout also was built in 1934 and, along with Easy

Ridge, was transferred to the National Park Service in 1969. The firewatcher at Copper—where the lookout is still in use—reported that the Easy Ridge building was leaning at a bad angle in 1972. It was burned the following year, with few to remember it. It was one of the least visited lookouts in the North Cascades.

The original trail was 20 miles long but logging roads reduced it by 6 miles before the roads stopped at the park boundary. The trail goes over Hannegan Pass and drops into the valley where the Chilliwack River begins. At 9.5 miles the river must be crossed without a bridge.

The trail has been officially abandoned so the junction isn't marked. The trail to Easy Ridge climbs a long, steep, and dry hillside to the lookout site. Glacier-clad Whatcom Peak and Mount Challenger are just beyond, on the east side.

Top, *Easy Ridge Lookout.* Bottom, *A 1910 photograph of Copper Mountain Lookout* (U.S. Forest Service photos)

Park Butte 🏠

Elevation 5,450 feet

The commanding view of Mount Baker, immediately to the north, is Park Butte's chief attraction. The Easton Glacier is so close that mountain climbers can be watched as they thread their way among the crevasses. The lookout, built in 1932, could have been put there as a "vista house."

Its elevation is not great but the drop-off between it and the slopes of Baker lends Park Butte the impression of being a rather high peak. Furthermore, its position, almost that of a shoulder of its massive neighbor, allows it an extensive view of taller mountains in other directions.

In the distance to the east are such 7,000-footers as Mount Blum, Hagan Mountain, and Bacon Peak. Closer by, in a west-to-southwest arc, are the orange cliffs of Twin Sisters Mountain—actually a small range in itself with half a dozen peaks ranging from 5,500 to 6,900 feet.

The lookout is leased to the Skagit Alpine Club of Mount Vernon, which maintains it. The club members have an annual work party, and the rest of

Park Butte Lookout and Easton Glacier on Mount Baker

Park Butte Lookout

the year Dr. Fred Darvill and his wife look after it. The cabin is open to the public on a first-come, first-served basis.

Dr. Darvill, then president of the Skagit Alpine Club, first visited Park Butte on Labor Day 1957. He remembers: "The lookout, Mike McGuire, hiked down to our camp in the meadow and asked us to come on up for the night. He hadn't seen anyone else for 39 days. He carried one of our packs to the lookout. I have been fascinated by lookouts since that time."

Logging on the south side of Mount Baker has continued to shorten the trail from its original 14 miles. In 1957 it was down to 9 miles and by now to 3.5, making it a popular destination for day hikers.

Packstring leaving Komo Kulshan Ranger Station for Dock Butte in 1936 (U.S. Forest Service photo by Fromme)

Dock Butte

Elevation 5,210 feet

From Park Butte you need no telephoto lens to fill the picture with Mount Baker. Now say you want to stand back a bit for a fuller view. You just go climb Dock Butte, 6 miles farther off and due south of Baker.

The butte stands on an east-to-west boundary of the national forest — everything north is federal, with state and private land on the other side. The lookout was supplied from the Komo Kulshan Guard Station, back when it was built in the mid-1930s. On the map the guard station was about 5 miles away, but by trail it was a dozen miles.

According to Dr. Fred Darvill of Mount Vernon, it was still a long walk in 1959, after the lookout had been abandoned. "The trail started about two-tenths of a mile inside the national forest boundary and went remorselessly upward. The lookout, a standard 14-by-14 with no balcony, was in a poor state of repair. I found its sign — giving the elevation as 5,300 feet — down on the trail and brought it home. I still have it."

Logging roads now make access much easier. A road intersects the old trail about 1.5 miles from the summit. Forest Service records say the building that once stood there was burned in 1964.

"There is burn debris left," Darvill reported, "so the site can be precisely identified. Telephone wire and insulators lie along the ridge."

On that occasion, he added, "For the first and only time in my life I found myself in the middle of the Mount Baker elk herd—in weather too dismal to permit photography."

Karst outcropping on the side of Dock Butte, with Mount Baker in the distance

Dillard Point

Elevation 2,400 feet

The lookout on Dillard Point was an undistinguished station, sitting on no more than a hill as elevations go in the northern Cascades. It has no historical significance. It was not built until 1962, when other lookouts were being abandoned, and it was surrounded even then by logging roads. Its site was chosen for an ugly view—fresh stumps. In 1975 the building housed electronic weather recording equipment, but it was vandalized and the building removed.

There had been a lookout on Sulphur Point, where the firewatcher could see into the lower Baker River Valley. It was reached by trail from the Komo Kulshan Guard Station. However, by the late 1950s logging activities had climbed to timberline on the slopes of Mount Baker. A road ran up Sulphur Creek, and the ridge between it and Dillard Creek was clearcut.

Sulphur Point Lookout was then abandoned and a new station was placed on a sturdy tower atop Dillard Point. There the lookout could keep an eye out for the fires that potentially go with logging and slashings, as well as watch the forest beyond.

The loggers receded into the distance, leaving nature to its reclamation project. Today a lookout at Dillard Point couldn't see the forests for the trees. He'd have to climb the tower to look over the top of the new growth before Mounts Baker, Shuksan, and Blum appear on the skyline.

Fog-covered Baker River Valley from Dillard Point Lookout

Bacon Peak and Watson Lakes on Anderson Butte

Anderson Butte

Elevation 5,420 feet

Man has changed portions of the view the lookout originally had from Anderson Butte. Back in 1936 when a building was first placed there, the fellow on duty must have had a tough time concentrating on his business instead of on the spectacular scenery all around.

Mount Baker dominates, but Mount Shuksan (9,127 feet) and Bacon Peak (7,066) are among the peaks that loom above all the others. Mount Watson, 6,324 feet, is only 1.5 miles to the southwest. Old Baker Lake was hidden from the lookout by a wooded shoulder, but he could see Anderson and Watson Lakes only a short distance away.

Two trails went to the top of Anderson, one from the Komo Kulshan Guard Station and the other from the Baker Lake Guard Station. Both involved climbs of almost 4,000 feet. Since then old Baker Lake, Baker Lake Guard Station, and 10 square miles of forest have been drowned in the reservoir created by Upper Baker Dam. Logging roads come within about 1.5 miles of the lookout site, and over on Mount Baker, clearcuts almost to timberline give its slopes a scraggly look.

Of course the peaks are still there for viewing. Although the building was removed in 1964, Anderson Butte became a vantage point for seeing the steam emitted by Baker when it started acting up in 1976. A vent reopened in Sherman Crater near its top, steam melted the surrounding ice, and there was concern that it might cause a mud slide that would end up in Baker Lake.

Sauk Mountain

Elevation 5,537 feet

Sauk Mountain, due north of Rockport, overlooks the junction of the Sauk and Skagit Rivers. The Skagit streams out to the west, down to Sedro Woolley. The view to the east continues along the Cascade River, which joins the Skagit at Marblemount. Sauk Lake is below the lookout, and the shortcut to it would be a 1,500-foot high-dive.

Once-forested Rinker Ridge lies to the south, and far beyond it Whitehorse Mountain can be seen. Southeast, forested hills lead up to Illabot Peaks. On the rest of the horizon are Eldorado Peak, the Pickets, Shuksan, and Baker. In late July the alpine meadows surrounding the peak are abloom with flowers, and the rocky summit on which the lookout stands is a natural garden.

The first building was put there in 1928. In 1957 it was replaced with a prefabricated building and tower, flown up in pieces by helicopter. The station was not much used, although it was maintained for emergencies until an epidemic of vandalism, and two burglaries of furnishings and the fire guard's personal belongings, led to the building's abandonment.

It was solid and could have been restored. However, the Forest Service does not have any money budgeted under the heading "sentimental expenses," and no volunteer group came forth, so the building was torn down.

A 1979 photograph of Sauk Mountain Lookout

Original Sauk Mountain Lookout (U.S. Forest Service photo)

When the first lookout was built, the trail started at the river level, but since 1950 loggers have worked their way upward. By 1957 a logging road reached timberline, leaving only about 1.5 miles of trail. It is broad and gentle but it does climb 1,700 feet up an alpine meadow that is dangerously steep in places.

Packing building material to Lookout Mountain, September 1929. As most of the mountain is forested, the picture was probably taken on top. (U.S. Forest Service photo)

Lookout Mountain

Elevation 5,719 feet

Lookout Mountain, just a mile from the boundary of North Cascades National Park, has a high rounded dome with a view up the Cascade River toward Cascade Pass, and northwest to Bacon Peak. Other glaciered peaks in sight include Big, Middle, and Little Devils of Teebone Ridge, and Eldorado Peak. A shoulder of Lookout blocks the view of the town of Marblemount and the Skagit River.

The building erected on the mountain in 1929 looked as though it had been put up by four carpenters who weren't speaking to each other. Each stubbornly followed his own plan, regardless of where that made his wall and roof join in with the other three plans. Perhaps the carpenters then assessed the result of their efforts and came to an agreement—that if the thing was to be a lookout, it lacked something. In a final cooperative endeavor, they placed an outsized doghouse on top as a cupola.

The present lookout, erected in 1962 with the aid of a helicopter, was certainly a contrast, one made more evident because its predecessor was left standing alongside as a storage shed. The new one is modern, with a flat roof, and is mounted on a tower of stout timbers. It is currently on standby status for emergencies. Maintenance work was done on the building in 1995. The Forest Service is looking for volunteers to continue its upkeep.

Unfortunately, one man's historic artifact is another man's eyesore. Without realizing its historic significance, a diligent forest crew in 1967 tore down and neatly disposed of the original building. If it had been kept, its age would now qualify it for the National Historic Registry, as oldest and oddest-looking lookout in the Mount Baker–Snoqualmie National Forest.

*Lookout Mountain. On the horizon are, left to right, Eldorado Peak, The Triad, and
Hidden Lake Peak.*

Hidden Lake Peak 🔭

Elevation 6,890 feet

The name is innocuous enough to be misleading. The scene suggested by "Hidden Lake" is a not-easily-discovered pond in the forest. In truth, while there is no trail to Hidden Lake, it is a relatively large and conspicuous body of water in a more-than-mile-high basin just inside North Cascades National Park. The peaks to the west and north are on the park boundary and rise to nearly 7,000 feet. The lookout, built in 1931 at the 6,850-foot level, is still in existence.

Hidden Lake Peak was the Forest Service's eastern outpost up the Cascade River, with a view to Cascade Pass. The next station beyond the pass, looking west up the Stehekin River, was McGregor Mountain. In contrast to McGregor, Hidden Lake was relatively easy to reach because of the old Cascade River Road.

The trail to the lookout, starting upstream from Marble Creek crossing, was 7 miles long. Horses made it to the saddle below the pinnacle, leaving half a mile to backpack, with a climb of 400 feet. A logging road has shortened the distance to about 4 miles of trail.

Lookout Margaret St. Aubin melting snow for water (Photo by Earl St. Aubin)

Earl St. Aubin and Bonnie in 1953. Below is the cloud-filled Cascade River Valley with Glacier Peak in the distance. (Photo by Margaret St. Aubin)

Earl St. Aubin and his wife, Margaret, were the lookouts in 1953. They were accompanied by Bonnie, a mixture of Irish setter and German shepherd who assisted them by proudly carrying 12-to-15-pound packs up that final 400 feet. In half a dozen trips, her total saved Earl a couple of backpacking loads.

St. Aubin recalls that it was a wet summer—there was a 4-to- 6-inch snowfall in August—and the only fires to report were from lightning strikes that turned some trees into flaming torches. He also remembers that bread could be baked at 6,850 feet, but beans were indeed *pièces de résistance*.

At the beginning of the 1960s the abandoned lookout was leased by the Skagit Alpine Club of Mount Vernon, under a special use permit. The club gave it up after a dozen years, but a small group of members, headed by Dr. Fred Darvill, continues to maintain it. The group calls itself Friends of the Hidden Lake Lookout, Lois Webster Memorial Shelter, after a member of the group who died.

Depending entirely on volunteers for labor and money, they re-roofed and painted the building, replaced broken glass, and put the stove into working order.

Lookout Barney Douglass at Devils Dome. Left, Fish caught in the Skagit River on his day off. Right, A visit by Lillian Elliot, his future wife. The photo was taken by her mother (and chaperone).

Devils Dome and Desolation Peak

Elevations 6,982 and 6,102 feet

Two lookouts who expected few or no visitors all summer were those on Devils Dome, now in the Pasayten Wilderness, and on Desolation Peak, 6 miles south of the Canadian border in today's Ross Lake National Recreation Area. Both sites looked into the Skagit Ranger District of Mount Baker National Forest. The district ranger was Tommy Thompson, a real original who was with the Bureau of Forestry when it became the U.S. Forest Service in 1907.

Both sites were also among the hardest lookouts to reach in Washington. After railroad and boat trips along the Skagit to the present location of Ross Dam, it was still a 20-mile hike farther up the river and into the mountains to get to either peak. So whatever supplies lookout personnel needed for the summer had to be planned for the first and only packtrip. If they ran out of anything, all they could do was call in to the ranger station in the faint hope that someone might have business up their way and bring it along.

For Barney Douglass, who spent two summers at Desolation Peak and

Jack Mountain from Devils Dome

two more at Devils Dome, one of his needs arose just because somebody did drop by. He was away working on the trail when a couple of young women arrived at the lookout. In a burst of sympathy for the poor lonely man who lived there, they whipped up a batch of fudge for him. The fudge was good, Barney concedes, but he found they had used up his entire summer stock of chocolate and sugar.

Desolation Peak was one of the many lookouts whose telephone line was subject to being knocked out by lightning. It was not always easy to locate

Ed Willgress, the lookout, photographed this deer at Desolation Peak in 1939. A large herd of deer lived in the Skagit River Valley and during the summer the bucks roamed the ridge tops. When the Skagit was flooded by Ross Dam, the herd lost its wintering grounds and died off.

the break in insulated wire, but Douglass described one method used if anybody came along to help. You had him crank the telephone every 5 seconds while he went down along the line, pinching into the wire. You knew the break was somewhere between you and the lookout cabin when you stopped getting a shock.

The Devils Dome Lookout was built in 1935, with a better view of mountains than of forest. By walking across the dome, one can still look down into the Skagit Valley and slopes to the west. To the south across Devils Creek is 8,928-foot Jack Mountain, which supports several respectable glaciers. On the horizon are Mount Baker, the Pickets, and so many other peaks identification is difficult.

To the south and east of Jack Mountain are the twin peaks of Crater Mountain, both of which once held lookouts. Neither one was popular with its human occupants. The peaks were subject to violent lightning storms, and the higher of the two, at 8,128 feet, had to have its materials and supplies hauled up by cable.

In the Skagit Valley, Ross Dam has flooded the remote waterfalls, and the largest trees have been logged off. The trailhead to Devils Dome is now reached by boat, but it is still 8 steep miles from the lake. Another trail runs 11 miles from Devils Dome to Holman Pass on the Pacific Crest Trail.

Desolation Peak is maintained for emergency use by the National Park Service. Among its past lookouts was the writer Jack Kerouac, whose book *Desolation Angels* describes his 1956 summer there.

North Mountain

Elevation 3,956 feet

North Mountain Lookout was something of an anachronism at the time it was added to the system. Other stations were being abandoned when the first lookout to stay on the mountain all season set up camp in a tent with a wooden frame. The year was 1962.

The site was a small clearing with a view limited by virgin forest. In 1966 a 14-by-14 building on a 40-foot tower was constructed to rise above it all. Since then, the top of North Mountain has been clearcut, which didn't improve the immediate scenery but does give a sweeping view of logging operations. In addition to miles of checkerboard patches, Whitehorse Mountain, Higgins, Baker, Glacier Peak, White Chuck, and Pugh are part of the panorama. On clear days the Picket Range is visible.

The lookout may have started as a laggard in history but as the main fire detection point near Darrington, the ranger district headquarters, it developed into a modern example of stations still manned throughout fire season. It even has a solar panel on its roof to furnish electricity for the radio.

North Mountain is maintained for emergency use. This dramatic viewpoint, easily reached by road, has been popular with the public. Unfortunately it has also been popular with vandals. To protect the building from further damage, the Forest Service has closed the road 1 mile from the lookout location.

North Mountain Lookout in 1979. Note the solar panel on the roof.

Mount Higgins

Elevation 4,849 feet

Higgins is a long mountain of 5,190-foot elevation on the western edge of the Cascade Range. The lookout, placed at the brink of a vertical cliff on a 4,849-foot promontory, offers a view of mountains on three sides and the Puget Sound country to the west, with the North Fork Stillaguamish River 4,500 feet below.

The original route to Higgins started with a difficult ford of the Stillaguamish on horseback. The trail that continued to the lookout, about 6.5 miles from the river crossing, was very steep, climbing about 1,200 feet a mile. It evidently was hacked out by miners in a hurry to get somewhere.

One of the miners was Sam Strom, a Darrington character who had a number of claims in the area. He neatly engraved a rock above the trail, "S. Strom, 8—1917." So, at least that part of the trail was already there when the Forest Service first placed a "rag house" on Higgins in 1918.

A building was not erected until 1926, but thereafter it served long and well. In the mid-1930s a lookout named Whalen went out at night to hunt for three boys reported lost on the mountain. He found them in a state of

Lookout tent, September 15, 1918 (U.S. Forest Service photo)

Remains of Mount Higgins Lookout, 1979

hypothermia, so he brought them to his station and stoked the stove so red hot it set the roof on fire. He duly reported the blaze the next morning and was given credit for a record-time initial sighting and extinction.

Another man on Higgins objected to one of his visitors when a bear climbed the steps and prowled around the catwalk, peering in the windows. The lookout phoned a complaint to the ranger station and was told, "Well, poke him off of there." Those orders were declined, and instead of the bear going over the mountain, the lookout did.

From the valley bottom, the south side of Higgins is very impressive, with steeply tilted layers of rock. The trail climbs the north side where the slopes are more reasonable. At the lookout site one has the feeling of sitting on an overhang, and the Stillaguamish below looks just a stone's throw away.

The lookout building was abandoned in 1965 and soon afterwards collapsed under heavy snow. Its remains are still there, thirty years later. Improvements in 1990 have made the old logging grades drivable, and the trail is 2.5 miles shorter. It is in good shape but as steep as ever.

Glacier Peak from Green Mountain

Green Mountain

Elevation 6,500 feet

Green Mountain, north of the Suiattle River on the western boundary of Glacier Peak Wilderness Area, has been used by lookouts since 1919. The building was not put up until 1931, and remained in active service until 1980.

Besides views of the Suiattle Valley, one can see Glacier Peak, Dome Peak, the red rock of Mount Buckindy, Mount Baker, Mount Shuksan, the Pickets, and even a small corner of Mount Rainier.

Green Mountain gets its name from huge green meadows visible for miles.

For all the roads that have been built in the past forty years, the lookout is still relatively remote, thanks to the 3,000-foot climb up to it.

John E. Schwartz, who in 1977 was living in Bend, Oregon, described in a letter conditions when he was the lookout in 1928. There was no road up the Suiattle and supplies had to be packed more than 20 miles, starting from a Sauk River crossing. Schwartz camped below the summit, in a patch of subalpine firs. He wrote:

"Lightbulb" Winders painting the lookout

> *My living quarters were an 8-by-10 wall tent, with a fly extending out in front. I cooked over a campfire, using a World War I army mess kit. There were no permanent improvements at the summit lookout site. An emergency ground wire telephone line was strung on trees and bushes along the trail to the top, where a small firefinder was positioned on a rock. There was no shelter and no material to build one, so it was necessary to spend long hours each day sitting on a rock in the sun.*
>
> *I carried my portable telephone—World War I surplus—back and forth between camp and the lookout so I was always ready for calls, in or out. Water came from the snowbanks on the northeast side of the ridge. Firewood was easily obtained from dead material in nearby thickets of fir and hemlock.*
>
> *During my stay on the mountain I saw two sheepherders, eight men in a Geological Survey party, and perhaps half a dozen Forest Service employees. When I hiked up the peak in August 1975, I encountered more people on that single day than in the whole season of 1928.*
>
> *A band of sheep trailed in from the Wenatchee Forest grazed on Green Mountain from late July to the latter part of August. The herders camped at the small lake in the basin just south of the main peak.*

The lookout was manned in 1975, 1976, and 1979 by Dick "Lightbulb" Winders. He gave the building a new coat of paint, and, with loving care, painted both sides of the wooden frames around 180 panes of glass.

In 1995 the building was condemned due to a rotting porch and problems with the underpinnings. Repairing the building is a high priority in the Darrington District. However, funding is a problem and the Forest Service may have to rely on volunteers and money from private sources.

Circle Peak with Mount Baker in the distance

Circle Peak

Elevation 5,983 feet

In 1930 lookout Jim Lewes lived in a tent until the building was finished. Once completed, the gable-roofed 14-by-14-foot "grange hall" built on a 500-foot pinnacle occupied most of the available space at the top of Circle Peak. The next resident was Andy Hollans, who manned the lookout in 1931. Not many visitors called in person, but the lookout could feel he had company when he trained his glasses around the skyline. He could see lookout stations on Pugh Mountain, Lime Mountain, Green Mountain, Huckleberry Mountain, and far in the distance, Mount Higgins.

Down along the Suiattle River were a few Sauk Indian cabins and the Forest Service's Green Mountain Pasture. To the east, 1,200 feet below, the sparkling waters of Indigo Lake must have been a temptation on hot days. Teamed with Green Mountain, cross-sightings could be taken up and down the Suiattle, and with Pugh on the south, there was a similar coverage of the White Chuck Valley.

When the station was built, lookouts headed for Circle Peak could travel part of the way by truck on a sketchy road up the Suiattle. At first they had to cross the river in an Indian dugout to reach the trailhead, but later a bridge was constructed.

The Sauk Indians no longer live in the valley although they still own some parcels of land, including a cemetery. Logging roads switchback up some of the hillsides. When the lookout was burned in 1967, the trail was abandoned. Although the trail has had no maintenance in twenty-five years, it is in remarkably good condition.

Miners Ridge

Elevation 6,210 feet

Although this is one of the most remote lookouts left in the state—16 miles from the nearest road-end—it has an unusual number of visitors. Some 3,000 persons a year reach the area on the nearby Pacific Crest Trail or hike up the Suiattle River to Image Lake, just east of the lookout. A few come by horseback, but the vast majority do it on their own feet.

Miners Ridge always has been in the heart of a wilderness, but it became a charter member of areas officially given that status when President Lyndon Johnson signed the Wilderness Act in 1964. The mining claims were donated to the county, leaving Miners Ridge with an uncertain future.

The lookout, built in 1938, is now used by backcountry rangers. It sits on a 40-foot tower standing on a grassy knoll. Around three-quarters of the compass are peaks extending into hazy distances. To the south, though, much of the view is blocked by 10,541-foot Glacier Peak.

No great parade of hikers reached the ridge in 1949, when Ira Spring photographed Mike Ohrberg going about his lookout duties. Mike's day started with the first light of dawn, because it is sometimes easier to see smoke before the sun comes up.

Lookout Mike Ohrberg using the firefinder

At 7:00 A.M. he reported by radio to the Darrington Ranger Station. Household chores included washing windows. Accompanied by his dog, Bounce, he descended the tower and with a saw and ax, hiked off the ridge to a forest a quarter-mile away, where he cut wood and packed it back. At 11:30 A.M. he made a second fire watch and at noon reported to the ranger station.

After lunch he went for water, which in early season he got by cutting chunks of snow from a nearby snowfield. In August, when the snow had gone and the ground was covered with flowers, he hiked a mile to the nearest stream to fill a five-gallon milk can—a forty-five-pound load. At 5:00 P.M. he checked again for telltale smoke and at 5:30 P.M. called Darrington with the day's final report. When the sun set he checked once more for smoke, then wrote up the report of his day's activities. In bad weather he spent all day working on trails, up to 5 miles from the lookout.

Mike Ohrberg and Bounce

Aerial view of Miners Ridge Lookout and Glacier Peak, 1968

The setting for his job was quite a switch from what young Mike had been used to. He was from South Dakota.

Considering the precarious position of so many lookout buildings, it was ironic that a serious accident, so rare elsewhere, should happen at this relatively benign spot. In August 1951 the Suiattle district ranger fell through the trapdoor entrance from tower to cabin and plummeted 40 feet to the ground.

A helicopter was called to lift him to a hospital. It crashed on the ridge. The pilot escaped but the craft was disabled, and the injured ranger had to be carried out by stretcher.

Dan Creek

Approximate elevation 2,000 feet

The Dan Creek station was a cabin in a recent clearcut, with a firefinder mounted on a stump. Whatever was visible from there has long since been obscured by new forest. The station is included in this book because it was representative of the many guard posts and temporary lookouts that backed up those at high levels. Their existence was barely noted, if at all, in permanent records.

Oblivion had overtaken this one, too, until builders of new logging roads, like archaeologists in a Mayan jungle, came across the ruins of ancient Dan Creek. The stout log framework of the building had been worked on by rodents and woodpeckers, and trees 10 inches thick now crowd around it. A number of artifacts—a rusted stove, tin cans, and a five-gallon water can— were discovered.

Thanks to a few stray pages unearthed from files at the Darrington Ranger

Lookout reports from Dan Creek and Pugh Mountain

Dan Creek Guard Station in 1979

Station, a reference to the water can came to light—along with a glimpse of activities at long-lost Dan Creek.

On July 18, 1941, the fire report reproduced here was made by the guard, Jim Tucker, to District Ranger Charles Thurston in Darrington. As it happened, the Pugh Mountain lookout already had forwarded the same news—that lightning had hit a large cedar snag across Dan Creek and it was burning, causing several other hot spots on the ridge. Although no other reports followed, it is probably safe to assume the fire is out by now.

At the end of the 1943 season the guard, George W. Rhoads, had left this friendly memo for his successor:

> *The firefinder fits into the top of a stump about 50 feet directly behind the shelter. The antenna pole may be used again. The garbage pit cover can be moved to a new hole when the present one becomes full. I left an assortment of grub you may use, if it is still good. It is in the big box by the writing desk. There is a full can of Velvet tobacco in the box, too; use it if you want to.*
>
> *The five-gallon water can had a mouse in it. It was bailed out but the memory lingers on.*

Apparently there was no guard in 1944 to pick up the note. Or maybe he read the comment about the water supply and applied for some other job.

Mount Pilchuck

Elevation 5,324 feet

When they first chose lookout sites in the Verlot Ranger District, Mount Pilchuck was a natural place to start. A mile-high peak on the western edge of the Cascades and the Mount Baker National Forest, it looked down on the North Fork Stillaguamish on one side and Pilchuck River along the other; and without peaks of any size to block the southern view, it looked clear to the Skykomish River Valley.

The first lookout was built in 1918 through the herculean efforts of those days, with materials hand-winched up rocky cliffs. The building probably was rebuilt later, because a 1939 photograph shows a well-preserved cupola of the 1920s period. It was replaced by the present hip-roof, which in 1960 was transferred to the state when the adjoining area became a state park. In recent years the lookout has been cared for by the Everett branch of The Mountaineers; they have remodeled it a bit and tried to make it as vandal-proof as possible, and turned the building into an interpretive center with photographs and artifacts. One hundred five people spent a total of 10,000

Mount Pilchuck Lookout, 1979

A small tarn near the top of Mount Pilchuck

hours on the project. The Snohomish County Search and Rescue and Army Reserve helicopters provided logistical support.

As late as 1950 the lookout was reached by trail from the Mountain Loop Highway in the valley bottom, a distance of about 10 miles. Logging and a ski development pushed the road upward to the 3,200-foot level so the trail is now only 2 miles long. The trail is used by hundreds of hikers every week; the foot traffic took its toll on the tread. The trail was rebuilt in 1995.

Hikers climb Pilchuck to see the whole Cascade Range from Mount Baker to Mount Rainier, and to the west, Everett, Puget Sound, and the Olympics. It is about as mixed-bag a panorama as can be found anywhere, its fascination enhanced by the combination of mountains, urban areas, and saltwater all in sight at once.

Pilchuck was the subject of a book written by Harry Higman and Earl Larrison (*Pilchuck: The Life of a Mountain;* Superior Publishing, 1949). Partly as a result of the book, Washington State Parks made a special arrangement with the Forest Service for recreational use of the mountain.

Top left, *Ranger Bruckart using the Osborne Firefinder, September 11, 1918.* Top right, *Nels Bruseth with the heliograph, a signaling device, September 1918.* Bottom, *Winch used to haul supplies to the top of the mountain. White Chuck Mountain is in the distance.* (U.S. Forest Service photo)

Pugh Mountain

Elevation 7,201 feet

Whatever other fire-spotting vantage points were used in the Darrington Ranger District's early days, Pugh Mountain, looking north into the White Chuck Valley and northwest down the Sauk River, was a pioneer in the system. A lookout camped on top each summer starting in 1916, and a building was erected in 1921. When it was junked in 1965, an era really ended.

Photographs taken in 1918 show the lookout living in a tent, with a tele-scope, a firefinder, and a heliograph as his equipment. The young man was Nels Bruseth, who during the rest of his life was so long associated with the Forest Service and local history he came to be regarded as "Mr. Darrington."

The lookout was on an airy perch amid a circle of peaks that includes Sloan, Glacier Peak, Dome Peak, Eldorado, Shuksan, Baker, White Chuck, Three Fingers, and those in the Monte Cristo region. William Pugh, a

Remains of the upper winch on Pugh Mountain, 1979

Pugh Mountain Lookout (U.S. Forest Service photo)

homesteader at the foot of the mountain at the beginning of the century, lent his name to it.

A horse trail extended from the valley bottom to the 4,900-foot level. From there a tortuous footpath led to the top. Supplies were hauled up on a two-section, hand-powered tramway, one section extending to 6,200 feet and the second one to the summit. The tramway was hung on a steel wire and pulled with a winch.

A logging road has shortened the climb by 700 feet, but that leaves a sturdy 5,300 feet to gain in elevation. The trail probably is about the same as when the lookout was there, including a few feet of knife ridge to traverse with a 200-foot drop on one side and a 1,500-foot drop on the other. The ascent ends with a scramble up a cliff, aided by footholds that were blasted in the rock.

Only the tie-down cables, glass, nails, bolts, hinges, and wire remain at the lookout site. The outhouse that stood in 1979 is now a pile of rocks. In spite of the rugged ascent, over 100 people a year sign the summit register.

Three Fingers

Elevation 6,854 feet

On a snowy September 26, 1929, Harold Engles, Darrington district ranger, and Harry Bedal, one of the district's near-legendary characters, climbed the south peak of Three Fingers. It was their only way to find out if a lookout could be put there, since nobody else had been to the top.

If anyone had scaled it before, the conclusion already would have been of record. No, a lookout on Three Fingers was out of the question. Fifteen miles to pack, with an ascension—not counting ups and downs—of more than 4,000 feet from the nearest road, and a treacherous permanent icefield to cross. The top 15 feet of the pinnacle would need blasting to provide room for a building, which then would need cables to hold it in place and ladders to reach it by other than rock-climbing techniques.

Three ladders are used to reach Three Fingers Lookout.

Aerial view of Three Fingers Lookout, 1968

None of those handicaps struck the Engles-Bedal team as impossible. From Three Fingers a lookout could see for hundreds of miles on a clear day—from British Columbia peaks to the tip of Mount Hood in Oregon, and out west to the Olympics, the Strait of Juan de Fuca, and Vancouver Island. The only nearby peak of comparable height was Whitehorse, 3 miles due north and all of 2 feet lower.

So the trail was cut and the telephone line strung in 1930. The building was guyed in place in 1932, overhanging a sheer 1,000 feet on its north side and a flush couple of hundred on the south. A small ledge on the third side was reached by crawling out a window. The lookout was ready for occupancy in 1933.

Among its occupants from then on were pack rats, obviously megalomaniacs determined to prove there is no human habitation so remote that a pack rat cannot invade it.

The lookout in 1936 was 6-foot-4, 210-pound Harland Eastwood, whose feats in mountaineering, and later mountain rescue, were discussed by friends with amazed respect. Even without a right arm, lost in an accident, he was a remarkably deft woodsman.

Eastwood took his bride Catherine to the lookout. Although newly married couples at lookouts were far from unique, the setting of high, remote Three Fingers made theirs a real honeymoon in the sky. As a result, Catherine's account of it was told the following year in the *Saturday Evening Post.*

The lookout in 1937 was Bob Craig, who moved over from a previous season on Mount Higgins. He shared the cabin with a dog named Perry. "Perro, Spanish for dog," he explained, but it turned out Perry didn't understand Spanish and his name was Americanized. Perry was of no particular breed, but watching him trot around on the ledges up there we thought any dog living on Three Fingers ought to be either an Airedale or a Skye terrier.

This classic lookout was abandoned too, but everything about it remains as before, thanks to the Everett branch of The Mountaineers and other volunteers. Not only did they repair rotting boards and broken windows, they had three 4-by-4 ladders built and weather-treated by shipwright Steve Phillips, then installed them at the top for the final ascent.

Harry Bedal and Harold Engles died many years ago. In August 1979, half a century after he and Harry first inched their way up the rocks to the top of Three Fingers, Engles hiked there again with friends, to celebrate his 77th birthday.

The building is now open to the public. As there is considerable exposure, those using the cabin and ladder must determine for themselves if the facility is safe.

Remains of telephone wire to the lookout. The summit of the mountain is in the distance. The lookout was located on the center peak.

Bench Mark Mountain Lookout (U.S. Forest Service photo by Pickett)

Bench Mark Mountain

Elevation 5,816 feet

An untypical cupola lookout which might have been one of four of its kind in the United States once stood on the highest point of West Cady Ridge. Bench Mark is a large, rounded mountain whose top is surrounded by fields of heather, flowers, and blueberries. Views are into the forested valleys of West Cady Creek and the North Fork of the Skykomish, with snowcapped mountains such as Sloan Peak, Monte Cristo, Baker, Rainier, and Three Fingers on beyond.

The lookout was built in the summers of 1929 and 1930. The frame was made from alpine hemlock found nearby. The siding and roofing were packed in by horse, over a 25-mile trail that started at Skykomish. If someone bound for the lookout was lucky, he could catch a speeder on the logging railroad and save 9 miles of hiking.

Norm McCausland, who became fire control officer in the Skykomish District, was the first lookout at the post. In addition to those duties, he patrolled all the way north to the Whitechuck Glacier and south to Evergreen Mountain. He was expected to check up on the miners in the area, and to make certain the sheepherders kept their animals in the proper grazing areas. He also was to convince them that killing all the bears they saw was a no-no.

McCausland once got mixed up in a quarrel between a shepherd and a packer. The sheepherder pulled a knife on the packer, and the packer responded with three shots from his rifle. Fortunately, he missed. The shepherd, wisely, withdrew for the time being, but a feud looked in the works. McCausland got word to Lake Wenatchee that all was not quiet on the western front, and help came in time to prevent bloodshed. McCausland earned his stripes as a "fire control" officer that day.

At first getting word out was not an easy matter. Before the Bench Mark telephone was installed, the lookout had to hike 3 miles to the nearest instrument at Cady Pass, where he could call the Lake Wenatchee Ranger Station. They relayed his messages to Skykomish, Bench Mark's district office. Later he had a direct line, but the 27 miles on a single wire was almost the limit of the day's technology. He had to crank long and hard to ring a bell in Skykomish, and then yell into the phone to be heard even faintly at the other end.

There is little evidence that a lookout building ever stood on Bench Mark Mountain. Some old stumps, a few telephone insulators tacked on trees, and a weathered signboard that had lost its messages remained until 1985. The long trail in has been shortened by logging roads, but it is still 8 miles from either the Skykomish or Little Wenatchee River Roads to the lookout site.

Bleached signpost on Bench Mark Mountain with Mount Rainier in the distance

Evergreen Mountain   🏠

Elevation 5,587 feet

"Named wrong" is the opinion of Ellis Ogilvie, who once served as the lookout on Evergreen. "In light of the mountain's history, it should have been called Everblack."

The lookout's job, Ogilvie said, "was to watch for the latest fire, set by loggers." The last big one was on the south side in the late 1960s, long after he had been there. Salvage operations stripped the fire-killed trees and that was followed by logging on the west and north sides. The lookout has not been used since the early 1980s.

When Ogilvie held the post there, another lookout existed at Surprise Mountain, far south on the Pacific Crest Trail. The schoolteacher holding that job for the summer said she could not get through to the Skykomish ranger station; there were too many peaks directly in the way. So would Ellis please relay the messages?

Ellis did, which of course entailed quite a bit of conversation with the young woman. In mock solemnity, he said, "I wasn't suspicious until too late." Another marriage among many was chalked up to a Forest Service long-distance introduction. The Ogilvies later served together on Evergreen when it was a wartime Aircraft Warning Station.

Much of the scene in the vicinity is devoted to timber as a harvest—an essential use, no matter if the immediate aftermath of cutting makes observers wince. Nevertheless, Evergreen presents long-range views of the Cascade crest peaks to the east, those around Dutch Miller Gap to the southeast, and the Monte Cristo peaks toward the northwest.

Logging roads used to salvage the trees killed in the 1960s fire shortened the trail to Evergreen to a steep 1.5 miles, but a flood closed the road in 1982, adding 7 miles to the trail.

Since 1990 the lookout has

White-headed woodpecker

Evergreen Mountain Lookout

been repaired and maintained by the Seattle Explorer Search and Rescue Team, Bill Rengsdorf, coordinator.

In 1990 the building was leaning 3 inches. From 1990 to 1995, some 4,600 man-and-youth days were spent repairing the building. Doors, shutters, and windows were replaced. The materials were supplied by the Everett Mountaineers and the Forest Service.

The Evergreen Mountain Road has been impassable since the floods of 1982. However, supplies and volunteers were flown in by the Army Reserve. The building is open to the public during the summer months.

View from Beckler Peak. Left to right: Mount Baring, Merchant and Gunn Peaks, Mount Townsend

Beckler Peak

Elevation 4,950 feet

Beckler Peak's high point is 5,062 feet but the lookout was placed on the shoulder of a false summit at 4,950 feet. The ridge bears a mixture of alpine trees and heather, interspersed with a few rocky crags. The view is north up the Beckler River to the Monté Cristo peaks and west to Baring Mountain. The town of Skykomish lies 4,000 feet below to the southwest, and due south is the Foss River Valley, with Mount Hinman, Mount Daniel, and the peaks above Dutch Miller Gap in the distance.

In the Skykomish Ranger District most logging, from its beginning, has been carried on in the vicinity of the Skykomish River Valley and tributaries to the north. The district's initial lookout was on Mount Cleveland, southwest of Skykomish. That peak is now inside the Alpine Lakes Wilderness Area, standing as one of its northern boundary markers.

The lookout station on Mount Cleveland was established sometime before 1920. It consisted of a tent, stove, cooking pots, and a firefinder mounted

Beckler Peak Lookouts. Top, *A 1924 photograph of tree platform and "rag house."* Bottom, *Cabin and lookout tower. Note how the original tree platform has been incorporated into the new tower.* (Photos by Norm McCausland)

on a wooden tripod. There was no telephone, so contact with the ranger station was by heliograph, using Morse code.

As roads were built, the fire prevention picture kept changing. In 1924 the equipment was moved to Beckler Peak, along with the tent. A lookout tower was devised by building a small platform between three living trees. Within a year a tower was erected and a log cabin about 10-by-16 feet was constructed of white fir. The workmen had to look hard for straight poles, and Norm McCausland remembered hauling some of them 800 feet by manpower. The lookout used the heliograph until a telephone line was strung to the ranger station.

At first the route to Beckler Peak was just straight up the mountain, but to get horses near the top a trail was built, switchbacking up the southwest ridge. (It continued to 5,200-foot Alpine Baldy, 3 air miles farther.) A tramway was installed on the last 500 feet of Beckler to carry water and heavy supplies. The lookout would fill a ten-gallon can of water, hook it onto a cable, then climb to the top and crank a windlass to pull up the load.

In 1950 the lookout was discontinued without premeditation. A winter storm snapped the guy wires and the tower toppled. The Forest Service retreated across the Skykomish River to Maloney Mountain, where they put up a 14-by-14 building that had been at Galena. It was cut in sections, packed up the mountain and reassembled. It stood until 1969, when it was burned.

The old trail to the Beckler Peak Lookout is completely lost under the slash of logging operations. By 1980, the only evidence left of the lookout was several telephone insulators still attached to a tree.

The 2700-acre Beckler River fire in 1935 was fought by 600 men employed by the CCC. (Photo by Norm McCausland)

Bare Mountain

Elevation 5,353 feet

Fishermen know the North Fork of the Snoqualmie River well. It flows through a vast unsettled region of lakes and tributary streams, made accessible by a maze of logging roads. All of it is state or private land except for that at the source of the river, at Prospectors Ridge, which is just inside the national forest. The ridge got its name honestly. It was poked, scratched, and staked from end to end, but the claims just promised without yielding much of anything.

Bare Mountain is the high point of the ridge, and the lookout there could see down into the Lenox Creek Valley on the southern side, and west into the upper North Fork's watershed. The building, erected in 1935, was removed in 1973, and the mountain acts as a northwest cornerpost for the Alpine Lakes Wilderness Area.

There were no smashing mountain views from the lookout, just miles of forest-covered hills with the tips of peaks showing down south around Snoqualmie Pass. Today, in three directions the forest is broken here and there with clearcuts. In the fourth direction, to the west on private land, entire ridges are stripped.

To the north is a rather disillusioning sight for anyone who knows Mount Index only by the stern, nearly vertical profile it shows to passersby on US Highway 2. From Bare Mountain you find that the impressive face is backed by a rounded south side only slightly higher than nearby unnamed hills. So Mount Index's reputation as a peak depends upon a sheer bluff.

A 5-mile trail up to Bare Mountain begins at Bear Creek, which under the circumstances should have been spelled Bare Creek. An abandoned miner's road follows the creek into the Alpine Lakes Wilderness, and the last 1.5 miles leaves the road and switchbacks up a steep meadow to the rocky crest.

Mount Rainier from Bare Mountain

Left, *Original lookout, September 5, 1920*. Right, *Second lookout under construction, July 16, 1924* (U.S. Forest Service photos)

Granite Mountain

Elevation 5,629 feet

Many of today's district rangers and their assistants were not yet born during the height of the building period, when the oldest lookouts were replaced by more modern structures and new ones were added to the system. Consequently quite a lot of pre-1930 history depends upon the memory of old-timers who were on the forests before paperwork was given much priority. They are likely to challenge dates and other information found in records that were compiled in later years, from whatever sources were then available.

A 1975 "Summary of Lookout Stations, Snoqualmie District" says a 12-by-12 with cupola was put on Granite Mountain in 1935 and that it was replaced by a 14-by-14 in 1956. But the cupola building actually was erected much earlier. It appears in a 1924 photograph. A photo dated 1920 shows that it was preceded by a cabin. So Granite must have held one of the district's earliest lookouts.

Both buildings were there on New Year's Day, 1926, when one of the authors (Byron Fish) and several other teenagers on Christmas vacation from high school stood on top of Granite Mountain. We took refuge in the cabin to eat lunch.

It is not likely that a new cupola building was put up in 1935—that style had become outmoded. Perhaps the old one underwent major repairs then, and was removed in 1956 in favor of the present building. Tourists have been trudging up Granite for the view of Rainier, Baker, Glacier, and the mass of peaks around Snoqualmie Pass from the time automobiles could reach the campground (and the summer cabins that soon followed) at Denny Creek. No one in our high school expedition yet had a driver's license or a car, so we got there by train. It let us off and later picked us up at a stop just west of the summit tunnel above Denny Creek.

The route to Granite's top is still a popular recreational trail, and summer visitors enjoy its views and extensive natural rock gardens along the way.

The lookout is now under the care of the Washington State Hi-Lakers, Gene Fraser, chairman. This group opens the building in May and closes it in October. With a minimum of three work trips a year, the building has been painted and some wood replaced. The Hi-Lakers also do some maintenance on the trail. Forest Service volunteers man the building during the summer months.

Hi-Lakers adopted Granite Mountain in 1985 and average about 250 hours on the building each year. One of the volunteers is Roger Schoenheals, who manned the lookout in 1957. The propane stove was in very poor shape so the chairman called all the repair shops before he found David Jocober, who was willing to make "house calls." Jocober made three "house calls" to Granite Mountain to repair the stove.

The third lookout on Granite Mountain

Stampede Pass

Elevation 3,963 feet

The lookout at Stampede Pass came late in the building boom of the 1930s, so it was not a historic site in comparison to many others. It sat on a knoll in the pass that was 1.5 miles wide, and it was reached by road. It looked around pragmatically at forested hills rather than at mountain scenery. The man on duty could see into the valley below Keechelus Lake and, on the west side, into the upper Green River Valley.

However, this lookout, along with its 35-foot tower, was destined to become a museum piece, an educational example of how such stations were built and what their human occupants did to protect the forests. It is now the tallest and heftiest visual aid used by Highline Public School District 401. It stands about 1.5 miles from the district's Camp Waskowitz near North Bend, 20 miles away from its original location.

The building and its tower are classical in that they are typical of an era. The hip-roofed 14-by-14 is equipped with an Osborne firefinder and weather instruments, and students are shown how they operate. There is a touch of "as it was in the old days" in the setup, but the basic principles taught—of

Opposite and above: *Stampede Pass lookout was airlifted and placed on a new tower at Camp Waskowitz in 1974.* (Photos by Ken White)

compass, map, and humidity readings, combustion, and the way land is divided into townships, ranges, and sections—are timeless.

At the North Bend Ranger Station, Ken White got word in 1974 that the unused Stampede Pass structure was to be destroyed. It was adjacent to a modern weather-reporting station whose crew could make fire checks if necessary. Still, one more lookout would disappear and White thought a few should be saved.

He enlisted the Air Force to move it, as a training exercise for big helicopters. A tower foundation was prepared at Waskowitz. The helicopter lifted the whole Stampede Pass cabin off the tower, flew to the new site, and placed it on the ground. Returning to Stampede Pass, the 'copter held the tower under strain while the legs were sawed free. Beneath the dust-storm blast of rotors at Waskowitz, workmen positioned the legs on the new foundation. The helicopter then picked up the cabin and replaced it on top.

The legs were shortened a bit by the chain saw surgery but the tower still rises 32 feet above a 900-foot ridge, so there is a panoramic view of the Snoqualmie Valley. Depending upon scheduled use (Waskowitz is busy all summer, with school programs and those of other organizations), the lookout is open to the public. A visit must be arranged in advance through the camp director.

Stampede Pass weather station near the site of the old lookout

Sun Top Lookout and cloud-capped Mount Rainier

Sun Top

Elevation 5,271 feet

In the White River Ranger District, Sun Top is the only lookout that is still maintained and manned—partly by Forest Service staff and partly by volunteers. It was reconstructed in 1990. Forest Service personnel keep it not just for its historic and sentimental value; it is a good station for watching thunderstorms. However, headquarters in Washington, D.C., demands justification for its cost, and on the home front vandals and thieves make maintenance difficult.

As a protective move, the ranger district uses no fire control money for the lookout, which was erected in 1933. The funds come from other departments such as public relations and recreation, so there are picnic tables and outhouses on the site.

The building is not on the tallest mountain in the area but it is strategically placed, a mile high to overlook a vast amount of forest land in the White River and Huckleberry Creek Valleys. To the north is Mount Stuart and Mount Baker. Mount Rainier, only 10 miles to the south, naturally dominates the scenery. Sun Top offers an excellent view of Rainier's Willis Wall, a 5,000-foot cliff, and of Winthrop Glacier.

This station, along with O'Farrell and Sawmill Ridge between Greenwater and Green River, was manned as an Aircraft Warning System post during the Second World War.

Originally the lookout was reached by trail, and a 1934 panoramic photograph shows horses packing in supplies. There was no road on a 1956 map but one eventually arrived. Now Sun Top is popular with visitors—including local residents—who drive up to watch the sun set on Mount Rainier. More than 5,000 people reach the lookout annually even though the road is steep and narrow and open only seven months out of the year.

Tiger Mountain

Elevation 3,004 feet

Logging went on in what are now the metropolitan areas around Seattle during the First World War and into the 1920s. In 1928 a forest fire swept across Tiger Mountain, in Issaquah's backyard. It is not properly on Mount Baker–Snoqualmie National Forest land, but Washington State had some lookouts of its own in this region, and Tiger became one of them.

The tower the state built on Tiger Mountain elevated it to a "peak" with a spectacular view, the tower being around 90 feet high. Arlene Brandal, the lookout in 1961, remembered the tower swaying even in a breeze: "Sitting up here is like being in a dinghy out in the Sound," she said. "I sleep like a log as the tower rocks me to sleep."

If a lookout wasn't in shape for mountain climbing at the beginning of a season, he or she was by the time it ended. There were 107 steps to the top of the tower; the instruments for humidity readings were at the bottom, and they had to be recorded four times a day.

Several lookout stations on the western fringe of the Cascades were within sight of Puget Sound and its cities, but Tiger Mountain had perhaps the greatest contrast in views. At night, 20 or more miles of Seattle's lights were strung along the landscape. In daytime downtown buildings and ferryboats could be seen, with the Olympics on the horizon. North to south, major peaks of the western Cascades were in sight, with Mount Rainier looming huge 50 miles away.

Tiger Mountain may be only a foothill of the big range, but it is still high enough to get an occasional mix of rain and snowflakes in early and late summer. Puget Sound's overcast, which can collect any time of year, was often just high enough to engulf the station. Arlene Brandal said that on a number of occasions, looking down from the top, she could not see the bottom of the tower.

Aerial view of Tiger Mountain Lookout and Mount Rainier, 1961

Lookout Arlene Brandal at Tiger Mountain Lookout in 1961. Top left, *Reporting to North Bend Ranger Station.* Top right, *Hauling up supplies.* Bottom left, *Relaxing at the end of the day.* Bottom right, *Recording lightning strikes during a late-night storm. She stands on an insulated stool.*

Like the lookouts on 7,000-foot peaks farther back in the range, Tiger Mountain was subject to full-scale lightning attacks, and the firewatcher had to stand on an insulated stool while recording the targets. Arlene Brandal's June 10, 1961 letter describes her first experience of a lightning storm, which happened also to be her first day of the job:

> *I was given instructions to pull the antenna plug when a storm was close and turn off the radio. Around 5:25 P.M. the antenna cord was hanging down towards the floor. Crash, snap, crackle, bang—everything at once. An arc flew out of the antenna about 3 or 4 feet towards the door. Needless to say, I was shook. The next day I started broadcasting like crazy as I was assigned to give the call letters for the district. Everyone was calling me and I'd answer, and they'd say, "no contact." Thought they were crazy and almost got on the air and told them so. Doggone it, I was here. At 10:00 A.M. a red truck came roaring up. As it turned out, lightning blew out my transmitter and five tubes in the radio, and they say this is the safest place to be! Since then lightning has also taken out the east-southeast window of my tower, and I still sleep like a log at night. But when there's a storm in the air, my hair is practically on end.*

Two weeks later she writes:

> *On June 16th . . . there was the biggest lightning storm I have ever seen. . . .There was one storm to the south and one to the east—heat lightning and strikes. The strikes looked like they were about 5 feet thick and just hung there. None of them were any closer than 12 miles or so, but the heat lightning lit up this place with an eerie blue-white light. All the surrounding hills were lit. I had my lightning stool out when the storm was still 35 miles away. . . .*

The station's only urban touches were that it was reached by road and the lookout had a gas stove to cook on. Otherwise, it was as remote as it gets. Another of Brandal's colorful letters explains:

> *I had [a] great thought the other night of swimming down in the spring about ¾ mile away, but a huge granddaddy bear beat me to it. The bear still had his winter coat and it had a reddish glow. He swam about 15 minutes and then wandered off down the road like he owned the place. Saw a huge doe and fawn also. I think the most fantastic sight I've seen so far was two huge bald eagles up here. I'm more at their flying level. They hung around for quite a while swooping and diving at something. Two grouse have also taken over the tower. The rooster gives out with his whoomp every once in awhile. . . .Gosh this life sure beats working in town. It's so nice and peaceful with a fabulous view and none of the rat race living of city life and no new clothes needed. I don't get lonesome. There is too much radio traffic and too much to see to really think about being a hermit. Wouldn't give up this job for anything.*

The lookout is gone now, replaced by a battery of electronic relay stations, and the mountain is crisscrossed by recreational vehicle tracks. The road has fallen into disrepair, but the views are still great.

2

MOUNT RAINIER NATIONAL PARK

The nation's fifth highest mountain outside of Alaska was a part of the original Pacific Forest Reserve, created in 1893. But Mount Rainier's unique stature as a huge volcano without surrounding peaks to lessen its impact earned it national park status in 1899.

Since much of the mountain is above timberline, it was the Forest Service, not the National Park Service, that took an interest in Rainier as a site for lookouts. It afforded outstanding views of Columbia National Forest (later Gifford Pinchot) to the south, and Snoqualmie to the north and east.

Gobblers Knob Lookout and Klantz Glacier. A telephoto from Mount Beljica

Anvil Rock

Elevation 9,584 feet

When the Forest Service chose Anvil Rock, a lava outcropping on the south side of the mountain, as the site for a lookout, it became the state's highest, though soon-to-come stations at Mount Adams and Mount St. Helens were higher. Built in 1916, the 9,584-foot Anvil Rock station peered down on the Tatoosh Range as though it were a ridge of foothills. In clear weather, the lookout could see far past Adams and St. Helens to Mount Hood and Mount Jefferson in Oregon. Early enthusiasm over this immense range of vision and the spotting of some fires that otherwise might have gone unseen, led to a proposal in 1920 to build the ultimate in lookouts—one on the 14,410-foot summit of Rainier itself.

Before that grandiose scheme could be put into action, however, someone made a closer study of Anvil Rock's five-year record. Clear days were few; the scene normally witnessed was either a snowstorm, enveloping clouds, or a vast, sunny, but impenetrable sea of clouds blanketing what lay below. The final conclusion was that the location actually was of little value in spotting fires.

However, the daily weather reports the station provided were so helpful

Left, *Second lookout building at Anvil Rock.* Right, *Wallace Meade, lookout from 1923 to 1938* (National Park Service photos)

Cornices on the Muir Glacier. Anvil Rock Lookout is located on the rock promontory at left center. Photographed in 1937.

Original rock lookout built in the 1920s. Cowlitz Glacier and Cathedral Rocks are in the background. (U.S. Forest Service photo)

to mountain climbers that the Park Service took over Anvil Rock in 1930 and manned it long after the Forest Service had given up on its other high stations at Adams and St. Helens.

The agencies changed but the human lookout didn't. From 1923 through 1938, when he was 66, Wallace Meade was on duty each summer. His presence was important to the site, apparently, because after he had gone, the building was abandoned and stood empty for a number of years.

Finally, about 1950, orders came from Washington, D.C. to remove the structure board by board and dispose of it outside the park. Evidently the head office did not understand the scope of such a chore, and no money was provided for a work crew.

One clear night when a thick layer of clouds covered the lowland country, "lightning"—or some such fortuitous disaster—struck the building. It caught fire and burned to the ground. The rangers were able to find only a few charred boards, which they dutifully put in sacks, carried down the mountain and, as ordered, disposed of outside the park. A March, 1992 letter from Dee Molenaar sheds a little more light on this story: "The Anvil Rock Lookout was deliberately/officially burned down by the park Rangers—George Senner, Forrest Johnson, and my brother K, in 1951—when it was above the Paradise cloud sea. They say it went very rapidly as the wood was very dry."

All that remains at Anvil Rock today is a level spot and a stone outhouse that has a tremendous view—down through two holes—of the Cowlitz Glacier several hundred feet below.

Other Rainier Lookouts

Elevations vary, 5,500 to 7,181 feet

After Anvil Rock's construction there was a gap of some years before more lookouts were built in Mount Rainier National Park. The next was at Colonnade, on the west side of the mountain, constructed in 1929 and opened in 1930. The lookout was built with materials packed in over 12 miles of trail to the site's 6,800-foot level. Howard Anderson of Packwood remembers it as one of those hard-to-reach, lightning-prone stations with a high turnover of personnel. "You'd pack in a lookout," he said, "and unless you hurried he'd beat you back out of there."

The lookouts apparently had reason for being leery of Colonnade. It was wrecked by lightning in the 1950s—or at least that was the official report. But it had not been used since 1931, when the lookout at Sunset Park 1,000 feet lower was added.

After Colonnade and Sunset came Shriner Peak Lookout to the east, at 5,834 feet. In 1932 a contractor packed in materials from the Chinook Pass highway to erect the lookout. It still stands, available for emergency use.

Gobblers Knob Lookout

Shriner Peak Lookout and Mount Rainier

Firewatchers had been stationed earlier at Tolmie Peak, Mount Fremont, Gobblers Knob, and Crystal Peak, but they lived in tents and were limited in equipment. Plans for buildings remained on the drawing boards until the New Deal arrived and launched public works projects. The CCC constructed bridges and trails, and in 1934 the Park Service put out bids for the lookout buildings.

The contract called for four hip-roof, two-story, 14-by-14 lookouts with encircling balconies. The lower room was to be used for storage. A Seattle firm built the structures; the Park Service made its own electrical installations; and the final bill came to $6,114. Three of them—Tolmie, Gobblers Knob, and Fremont—are still in use, manned in the summer primarily for ranger contact with hikers and climbers.

Tolmie, Gobblers Knob, and Fremont are on the National Historic Register.

3

THE OLYMPICS

The interior of the Olympic Peninsula was largely unexplored when it was put into the Forest Reserve in 1897. It was obvious even from the fringes, though, that the peninsula contained the thickest, most extensive stand of big timber in the United States.

The loggers already were aware of that. What was to become the climax of the Paul Bunyan era was gathering momentum on the south side of the peninsula. It was speeded up after the San Francisco earthquake and fires of 1906, which created a huge demand for lumber. Grays Harbor sawmills proliferated, and the harbor became the biggest lumber-shipping port in the

Miners Creek fire from Mount Jupiter (U.S. Forest Service photo)

Top left, *Finley Creek Lookout* (U.S. Forest Service photo). Top right, *Webb Lookout* (U.S. Forest Service photo). Bottom, *Hyas Lookout* (Photo by Jack Rooney)

world. By the 1920s the timber harvest ran to more than a billion board feet annually in Grays Harbor County alone.

In the rainy forests on the southwest side of the Olympics, reality sometimes matched the Bunyan legends. A case in point was the famous "21-9" (Township 21 North, Range 9 West) near Humptulips. Trees up to 14 feet in diameter were so close together the loggers barely found room to work and had to fell them all in the same direction. They started cutting in 1909; by 1936, 28 of the 36 square miles on private land had yielded more than two billion board feet, with as much yet to go. Township 21-9 set a Douglas-fir record never surpassed elsewhere.

As early as 1904 a Tacoma congressman, Francis W. Cushman, tried to establish an "Elk National Park" in the center of the Forest Reserve. His bill failed, but in 1909 Theodore Roosevelt proclaimed 615,000 acres to be Mount

Olympus National Monument. The renamed Olympic National Forest neatly framed it on all sides.

The difference between a monument and a national park didn't mean much to the public, and all but the most elderly Washingtonians "remember" that there always has been a "park" there. Actually, the change in status came in 1938, along with a considerable expansion in size. Additions were made in 1940 and 1943 before Olympic National Park was formally dedicated in 1946. Later the Queets Corridor and the Ocean Strip were added, all of which tends to complicate dates concerned with the park and its boundaries.

The park attained another type of designation, the first of its kind: "World Heritage." Olympic National Park is one of three U.S. nominations to be registered as an "International Heritage." (The other two are Mammoth Cave and the Wright Brothers National Memorial in North Carolina.)

The lengthy evolution of the park, now fifth in size in the nation, kept changing the borders of Olympic National Forest. Today the forest is a horseshoe around the park, thin and broken on the north with open ends to the west. Between the park and the Pacific Ocean lies the largest forest administered by Washington State.

With the heavy lowland timber on state and private land, and the high peaks, meadows, and wilderness scenery inside the national park, the Forest Service was left to watch over the in-between band. Its five districts are strung end-to-end along a 220-mile arc, the shortest driving distance if you were to visit each ranger district. Since the widest part of any of the curving belt is about 15 air miles, Olympic National Forest probably can claim to be the longest, skinniest national forest in the whole country.

Although it was a charter member in the Forest Service, Olympic National Forest appears to have been left somewhat to itself. Other national forests had common boundaries, from the northern Cascades into California and west-to-east as far as the Dakotas, but Olympic was off in a detached peninsular world.

Furthermore, its future shape was in doubt during the lookout-building boom of the 1930s. A national park much bigger than Olympic National Monument was proposed long before it actually came about. It was only natural that the Forest Service should first put its latest developments elsewhere, where they seemed certain to remain part of its own coordinated system.

Maybe those are the reasons why the Olympic National Forest seems to have had more do-it-yourself lookouts than the others. Nonstandard buildings ranged from Pete Miller's Treehouse to the tower on the Hoodsport District that looked like a square Dutch windmill. Webb Lookout's bottom floor served as a garage and woodshed, the middle floor as living quarters, and a large cube on top as the watch station. After this strange edifice was transferred to the state, it was torn down.

Another reason for the free-style architecture might be that a high proportion of the Olympic Forest lookouts were drive-ins. Highway 101 long has encircled most of the forest, allowing entrances from all directions. Many stations were built because a road already had tapped the area, and a lookout person was put there to keep an eye on logging operations.

While a lookout in the Cascades might not see a man-made mark on all the land he watched over, in the Olympics it was a different story. In the Cascades, parts for modular lookouts were packed in over as much as 30 miles of trail, and the station was to be a permanent fixture, guarding the future. In the narrow Olympic band, what happened was often a matter of the moment. Here, logging had already begun, and if a lookout quickly built from trucked-in lumber served the purpose, it could do for now.

Hyas, in the Forks Ranger District, is a ridge of about 3,000 feet in elevation between the Calawah and Sitkum Rivers. A makeshift two-story building was put there after a fire in 1950. It appears in a 1956 photograph, and foundations on the site indicate that a low tower may have been added later. However, the whole Hyas Ridge was crisscrossed by roads and was clearcut. It is now a tree farm, and new growth has all but hidden any view.

The first lookout in the Olympic National Forest was built in 1915, on a forested ridge of 3,419-foot Finley Peak. It was on the Finley Creek side, looking into the Quinault Valley and along Matheny Ridge. The log cabin with its pyramid roof appeared to have been designed by gnomes. It stood until 1947, by then in the national park. The Park Service removed the building; nature removed the trail to it and obliterated the historic site itself with tall trees and underbrush.

Higley Peak, elevation 3,025 feet, was just north of Lake Quinault at the park boundary. In this case, Olympic National Forest was dealt in on the new-model lookouts, and a standard hip-roof was placed on Higley in 1932. Even when other stations were being closed, it was remodeled in 1964, on a short tower. It lasted until 1973, when a large radio reflector took its place.

Higley was manned year-round during the Second World War as an Aircraft Warning System (AWS) station. Many other peninsula lookouts performed that duty too—the war years saw the peak of the lookout network on the Olympic National Forest as well as on state lands. Closest to the Pacific, they all were busy guarding against the possibility of arson by enemy raiders.

Near Blyn, the state modified its 90-foot wooden tower with a stairway up the inside, to make it an AWS station where women would be willing to stand watch. Originally, the lookout just clambered up a ladder on the outside.

On Mount Walker near Quilcene (another CCC-built drive-in with garage and storage shed), a family of four held the post. The husband was the lookout five days a week and the wife was hired for the other two days. That way, they couldn't draw overtime.

Dodger Point, elevation 5,753 feet, is the only lookout still standing in Olympic National Park. Built in the 1930s, the location is so remote and isolated that in recent years it has only been used by an occasional backcountry ranger and seldom sees hikers.

Sun rising directly behind Mount Rainier, casting a reverse shadow on high clouds

Capitol Peak

Elevation 2,658 feet

A rugged section not much known to the general public is 80,000-acre Capitol State Forest. It lies in the quadrangle between old Highway 99 on the east, US 410 between Olympia and Elma on the north, and State Highway 12 from Elma to Centralia on the west and south. Its "mountains" are the Black Hills, of which Capitol Peak is the highest.

The area is well-tapped by roads, some 75 miles of them, familiar mostly to what author Harvey Manning calls "razzers"—those who ride on motorbikes or in four-wheel-drive vehicles.

Capitol Peak was the site of an important state lookout, allied to the circle of state and federal observation points encircling the Olympic Peninsula. It is only 2,658 feet high, but commands a view from the Cascades to the Olympics, from the southern tip of Puget Sound to the Pacific Ocean, and far south over the hills and prairies bordering the Cowlitz River.

You can drive there from entrances on three sides. In common with a number of other peaks, Capitol has switched from lookout to electronics. A Martian-creature tower, with shields and antennae, now stands on the spot where the lookout station used to be.

Kloshe Nanich and North Point

Approximate elevations 3,000 and 3,400 feet

L ake Crescent and Sol Duc River Valley, through which Highway 101 runs, are blocked off from the Strait of Juan de Fuca to the north by a 19-mile-long ridge that rises to more than 3,700 feet. Since its base is not far above actual sea level, it forms quite a respectable mountain.

On the western half of the ridge is Kloshe Nanich, an Indian name meaning large view. The lookout was built in the late 1920s, in the cupola style of the period. The CCC built a road to the lookout in the 1930s.

Kloshe Nanich, elevation 3,000 feet, sat on a rocky shoulder on the south side of the ridge, with a superb view of Mount Olympus and the Sol Duc River Valley all the way westward to the ocean and eastward to Lake Crescent. It would have looked down on the blackened stumps of the giant Sol Duc burn of 1925. Kloshe Nanich was replaced in the late 1930s by North

A 1996 photograph of the rebuilt Kloshe Nanich lookout

Sunset over the Pacific Ocean from Kloshe Nanich Lookout

Point, located a mile further on the 3,400-foot ridge top. North Point's view of the valley wasn't quite as good, but it could watch over the north side of the ridge, a position that also gave it a wide view of the Strait of Juan de Fuca. Both lookouts had roads to them.

North Point Lookout was standing in 1979. However, it was in sorry shape from vandalism: shot up, broken into, and left with loose shutters that banged against the windows and smashed the glass before they were torn off by storms. It furnished a pointed example of why the Forest Service gave up trying to preserve its unattended lookout buildings, even for emergency shelters, and destroyed most of them. In the 1980s North Point was leased to the county and restored to house communication equipment.

Kloshe Nanich Lookout has been rebuilt as an interpretive center, using the original plans and the still-standing steel post that holds the firefinder. From the museum a trail leads eastward to Mount Muller.

The reconstruction was made possible by a grant from the ITT Rayonier Foundation and supporting grants from Washington State's Interagency Committee for Outdoor Recreation (IAC) and the Forest Service. At the same time, 15 miles of new trail from the valley bottom to the meadows on Mount Muller have been constructed.

Bogachiel Peak

Elevation 5,474 feet

Bogachiel Peak is the highest point on 5-mile-long High Divide, which separates the Hoh River from the Bogachiel and Sol Duc Rivers. From the lookout site the mountain, forest, and lake views are sensational, with Mount Olympus—directly south across the Hoh Valley—the dominant feature. On the north are five of the lakes of Seven Lakes Basin.

The Olympic marmot, a species found only in the Olympic Mountains, lives here, and during summer months when blueberries are ripe, bears are frequently seen. High Divide also abounds in wildflowers.

In 1950 a young couple, Bill and Mary Jane Brockman, spent their honeymoon at Bogachiel Peak Lookout. The Park Service lent them a horse to pack in their supplies, with the advice that when they unloaded, they had only to let the horse go and it would find its way home.

As they plodded up the Sol Duc Valley it began to rain, and by the time they reached the lookout, it was snowing and blowing. The couple opened the building, unpacked the horse, and, as instructed, gave it a slap to send it on its way.

The horse just stood there. When they went inside, it came and peered

Elk on the side of Bogachiel Peak

Mount Olympus and avalanche lilies from the trail to Bogachiel Peak

forlornly through the window. Snow was sticking to its mane and its back and the couple were afraid the woebegone creature would die of exposure. They brought it inside their 10-by-10-foot house for the night. When they phoned the ranger the next morning, they probably could report that conditions were stable.

Thanks to the protection of the national park, the views are the same as when the lookout was built back in the late 1920s. The building is long gone and the trail has been shortened by 2 miles, but otherwise the elk still graze near Bogachiel Peak, the flower fields bloom gorgeously, and bears still prowl campsites looking for unattended tents and packs to plunder.

Hurricane Hill Lookout (National Park Service photo)

Hurricane Hill

Elevation 5,757 feet

It began as a remote Forest Service lookout and became a tourist attraction for people willing to stroll a short 2 miles on an asphalt path. Those who don't take the walk are missing one of the most spectacular viewpoints in the state.

The lookout has been replaced by an interpretive center with a nature trail. Hurricane Hill looks across the chasm of the Elwha Valley into the heart of the Olympics. Below in the other direction is Port Angeles, the Strait of Juan de Fuca and, on beyond, Vancouver Island. On clear days you can see the San Juan Islands and Mount Baker.

Several trails led to the lookout from sea level, so it must have been there before the CCC bulldozed the hair-raising horseshoe turns up from Whiskey Bend, on the Elwha River, to Hurricane Ridge. The road went to Hurricane Hill on the north and 10 miles south along the ridge to the base of 6,450-foot Obstruction Peak. It probably was headed for Deer Park when the project ran out of money.

Just south of Hurricane Hill was a fantastic flower field known simply as "the Big Meadow." So everyone could see the flowers, a highway was built from Port Angeles in 1958, and thanks to its popularity, a lot of the Big Meadow was paved for parking and sidewalks—which made it harder for the flowers to grow there.

Fortunately the highway stopped 2 miles short of the lookout site, the old road was closed off, and the asphalt path takes over the rest of the way up the hill, with a climb of 700 feet.

Before the Olympics became a national park in 1938, the Washington State Game Department transplanted mountain goats to Hurricane Ridge in anticipation of making them a game animal. Without natural predators, and with hunters not allowed, the goats multiplied and spread until they have threatened the park's original ecology. These frequent and friendly visitors around the lookout site and other areas of the park are now subject to deportation when they exceed a population quota.

Unnamed meltwater lake below lookout site

Panorama photos taken June 21, 1935 show the effects of the Maynard Peak burn which took place early in the century. Note the lookout railing in the lower corners. (U.S. Forest Service photo by Sarlin)

Ned Hill

Elevation 3,450 feet

Paul Bunyan was a grown man before he came west. Otherwise, historians might conclude that when he was a small boy, he built Ned Hill Lookout. It is a platform on a makeshift tower of snags and poles. Some poles are placed on a diagonal from ground to top, to brace the framework. It must have taken a pretty husky kid to put them in that position.

No one knows how long the contrivance has been standing, other than that a panoramic photo was taken from it in 1935. The scene then was one of snags and small trees, far into the distance. A fire early in the century had swept all across the country around Maynard Peak and Ned Hill. If the platform was erected in the 1920s to keep watch over the burn, it may be the oldest existing lookout on the Olympic Peninsula.

During dry periods in the 1930s a fire guard was stationed on Slab Camp Creek, a stream to the southeast of Ned Hill. He went up to the platform on

the rounded dome of Ned Hill to look over the Gray Wolf River Valley and the slopes of Maynard and Baldy. Blue Mountain (site of Deer Park ski area) was 4 miles directly west, and to the north he could see the Strait of Juan de Fuca.

The forest has grown since the 1935 picture and the old platform is hidden by tall trees. The trail from Slab Camp is no longer needed by fire guards; having no special recreational value, the trail was abandoned for a number of years and was overgrown with salal and 10-foot-high rhododendrons. In 1994 volunteers opened the trail wide enough to wiggle through. The trail climbs 900 feet in a long mile, to the lookout tower and platform.

Lookout platform in 1978

Anderson Butte

Elevation 3,358 feet

The Anderson Butte Lookout was a CCC project in the mid-1930s. It was perched on a rocky outcropping, and building it resulted in a tragedy that was remarkably rare in the history of lookouts, considering the dangerous sites on which so many of them were placed. Here a construction worker was killed in a fall during the initial building of the lookout.

Anderson Butte is on the south side of the Olympic National Forest, squarely in the middle of the Simpson Timber Company's domain. The view is of the Wynoochee Valley, Camp Grisdale and, in recent years the Wynoochee reservoir. To the north at some distance, one can see the icefields of Mount Olympus.

When the lookout was removed, the trail was abandoned and much of it has been obliterated by logging. It is still shown on the Olympic National Forest recreation map but only the last few hundred feet of it is recognizable as a trail. To find it, one first must unravel the maze of logging roads and arrive at the right clearing.

A visit to Anderson Butte is certainly not a wilderness experience, but it has its place in the overall subject of lookouts and what they were set up to do—protect a commercial crop of trees from fire. The southern section of the Olympic National Forest is and always has been logging country, much of it now in the second cycle of sustained yield. The "highest and best use" of the land has been to produce a necessary product.

Courtship display of male blue grouse at Anderson Butte

Cable handrail to Anderson Butte Lookout in 1952. Part of the handrail could still be found in 1978. (U.S. Forest Service photo)

Pete Miller's Treehouse (Cook Creek)

Elevation 264 feet

In itself, there was nothing unusual about using a tree as a lookout tower. It was done all the time, especially in the days when few buildings had yet been erected and lookouts camped in a tent at or near an observation point. They took advantage of handy trees by attaching ladders to make ascent easier and sometimes building a platform on top where they could sit while scanning the forests.

The Cook Creek example, though, was in a class by itself. High-climber Pete Miller turned a fir into a 152-foot stump, and the 7-by-7 cubicle placed on top probably became the world's highest treehouse.

Miller cut side notches 4 feet below the top and spiked railroad ties into them for floor joists. A rigging crew-turned-carpenters laid planks across them and hoisted up pre-built walls. With the walls up and subject to breezes, the roofing job was made all the more entertaining by the swaying trunk. The part of the tree projecting into the cabin acted as a base for the firefinder. The spiral ladder up the length of the tree was made from some 180 steel rods driven into the trunk and joined together by logging cable.

As a final afterthought, a rigger scrambled out a window and onto the roof to set up a flagpole made from waterpipe.

This remarkable lookout was on the Quinault Indian Reservation, on a ridge above Cook Creek. It was used from 1927 until 1940, when a steel tower was built nearby to overlook the same general area.

The stripped tree stood for some years afterwards. A 1947 photograph showed it as a lonely spar rising a hundred feet above surrounding new growth. Apparently it succumbed at last to age and storm, for it is no more.

Steel lookout tower on Lone Mountain near Cook Creek

Pete Miller's Treehouse. Note the man standing on the spiral ladder in the photo at right. (U.S. Forest Service photos)

Remains of the lookout, with Mount Washington in the distance

Jefferson Ridge Point

Elevation 3,832 feet

This latecomer was built in 1961 on a rocky, tree-studded ridge between the Hamma Hamma River and Jefferson Creek Valley to watch over a logging operation. The trees were cut from the rounded knoll so the lookout had an imposing view. For half an arc the peaks in sight were Washington, Pershing, Sawtooths, Stone, Bretherton, and The Brothers. In the rest of the circle were Hood Canal, Puget Sound country, and the Cascades from Mount Adams to Mount Baker.

From the 2,000-foot level up, the forest floor is covered with rhododendrons 3 to 6 feet tall. They grow along the ridge and at the lookout site.

The lookout was probably the last one built in the Olympic National Forest, and the method of assembling it was prophetic. Its parts had been stored at Hoodsport, and during a forest fire in 1961, when a helicopter working on the fire was stationed at Hoodsport, the ranger had the material flown in. The lookout was used only a couple of years and was burned in 1967. It was no longer needed because helicopters could patrol the area.

The site is reached from the Hamma Hamma River by a very steep trail, climbing 2,800 feet in 3 miles. At times it follows tractor tracks around a clearcut so steep the tractor must have gone up and down with the aid of a cable.

4

GIFFORD PINCHOT NATIONAL FOREST

This region of southern Washington, part of the original Mount Rainier Forest Reserve of 1897, became the Columbia National Forest in 1908 and was renamed for forester Gifford Pinchot in 1949. It spreads from Mount Rainier National Park to within sight of the Columbia River, but unlike the Mount Baker–Snoqualmie National Forest, it has only one highway that passes through its 1.25 million acres.

Even that one, State Highway 14, ticks only the extreme northern end, following the Cowlitz River Valley until it takes off to climb over White Pass.

Sunrise on Mount Rainier from near Beljica

Sleeping Beauty, left and Mount Adams

From Randle or Packwood, north-to-south passage is possible—depending upon weather, road repairs, etc.—by a couple of tortuous routes. One comes out at Trout Lake in Klickitat County, and the other follows the Wind River Valley down to Carson on the Columbia.

That doesn't mean Gifford Pinchot is a roadless region. There is a maze of roads, especially in the southern part, all designed for logging. In places, their pattern looks like a cracked windshield—a not unlikely simile if you meet enough logging trucks. When the map shows an "all weather" road, it may recently have been improved with crushed rock put through a 3-inch screen. If your car wears logging truck tires, you'll enjoy the ride.

The Pacific Crest Trail manages to squeeze down the national forest's eastern line, much of it along the Yakima Indian Reservation boundary, until it comes to Mount Adams There the trail meekly yields right-of-way and detours to the west. Having passed the big hump, where does it go to stay on top? The only choice is along a ridge between the Little White River and the Wind River, which both flow south to the Columbia. The ridge offers strategic positions for lookouts, and the Crest Trail wanders through or near the sites of half a dozen of them.

One, Red Mountain, along with numerous cinder cones in the vicinity, is of volcanic origin. Below it to the east and south is Big Lava Bed, some 30 inhospitable square miles self-created as a roadless area. Geologists say the rock down underneath has not yet entirely cooled.

Red Mountain, reached by a road from Carson, is in the huckleberry belt. The highest sustained yield is around Mount Adams but the Klickitat and other Yakima Indians came here, too, to harvest the crop and play games, among them horse racing. Visible from the old lookout site are ruts in a meadow, showing where the track was.

Early lookout on Red Mountain. The caption on the back of the photo says "1910 photograph," but records indicate the lookout was not built until much later. (U.S. Forest Service photo)

In common with those of other national forests, Gifford Pinchot's map is sprinkled with the names of pioneer families who were, in effect, founding members of the Forest Service. In little towns—Morton, Randle, Packwood, Trout Lake—and back in cul-de-sac valleys where farms or small mills were started at the end of the last century or the early years of this one, second and third generations carry on the tradition of working for the Forest Service.

In Randle, Keenes Mead retired from Forest Service packing in 1969. His father-in-law, C. A. Gardner, built the lookout on Kiona Peak in 1918 and the one on Tatoosh soon afterwards. Tony Guler shows up in the records as an assembler of the Sleeping Beauty Lookout in 1931, but by then the name of Guler was long established as a guard station near Trout Lake. Tony was of a later generation. In Mineral, Jim Hale's father worked for the Forest Service long before Jim did.

Nevin McCullough's father homesteaded in 1890 near Ashford, where Nevin was born. When the Forest Service began, the senior McCullough was named "Forest Ranger" for the district. He served until 1921 and was the first in Washington State to retire from the Service.

Red Mountain Lookout and Mount Hood

Nevin himself started working for the Forest Service on the Naches District in 1922, with the usual jobs in summer and log scaling during the winter. In 1927 the ranger on the White River District was transferred to Naches and Nevin replaced him at $125 a month. "They chose to send me there," he says, "because I was the most dispensable." For the rest of his Forest Service career, until he retired in 1959, Nevin was White River's district ranger.

Nevin was early interested in aircraft, leasing an open-cockpit Stearman on standby in 1928. Planes were not taken very seriously, though, for about ten years. The improvement of radio helped bring them into their own. After the Second World War McCullough flew over fires, "talking smokechasers to the right place."

However, not all the names of lookouts came from human old-timers. There was no Pompey family to go with Pompey Peak, overlooking the Cowlitz from south of the river, midway between Randle and Packwood. Pompey was an old mule who packed stuff up there to build a lookout, lost his footing, and died in the line of duty.

Keenes Mead and Howard Anderson of Packwood both were packers for the Forest Service, but on different terms in adjoining ranger districts. A ranger

Loading a packstring

was free to own a herd of horses and mules (which meant taking care of them all year), or he could farm out the packing to private individuals at the best deal he could make. One packer might manage a dozen to a hundred Forest Service animals, supplying only his own riding horse, while in another district the packer provided them all on contract.

The big source of government mules was the Ninemile Remount Depot, 30 miles northwest of Missoula. It came into being because 1929 was such a bad fire year in Montana. The Forest Service drafted all the horses and mules it could round up—1,500 of them—and still was short of stock.

So the Remount Depot was begun in 1930. It was built to compound size in 1933 by the CCC, patterned after Army cavalry stations. The modern touch was specially built trucks that could rush loads of ten pack animals and their cargo to fires from eastern Washington to the Dakotas.

Starting with several hundred head, the Forest Service developed its own breeding ranch at Ninemile and a 44,000-acre winter range, 90 miles north of Missoula. The depot was operated for thirty-two years, until forest roads, aircraft, and the first smokejumper base (also at Missoula) led to a decision that the packstring factory no longer was needed.

It didn't greatly affect western Washington, except that a district ranger always knew where he could get his own mules if he wanted them. Nevertheless, Ninemile Remount Depot held a distinctive position in Forest Service annals. Thanks to district ranger Jerry Covault and his wife, the compound has been restored. On July 19, 1980, it was dedicated as a National Historic Site.

Mount Beljica

Elevation 5,475 feet

This was a surveillance spot with no building. The lookout had a close-up view of the west side of Mount Rainier, and also could see down into the Nisqually Valley and out to Mount Adams and Mount St. Helens. Other lookouts in sight were on High Rock, Glacier View, and, inside the national park boundary, Gobblers Knob, close enough that anyone at Beljica could tell whether its windows were open or shuttered.

Beljica was reached by trail from near Ashford or through the park. It had a telephone and a small toolshed 100 feet below the top. The nearest cabin, which Nevin McCullough helped build in 1922, was down at Lake Christine

Summit of Mount Beljica

A 1979 photo of Mount Beljica toolshed and the board where the telephone was attached

half a mile away. After a lightning storm, a Forest Service employee would go up on top for a look around.

This type of station was not unusual. It and others of its kind were sort of non-coms in the system, ranking a telephone but not a man posted there with the title of lookout. They were used by whoever was assigned to the area as fire warden, patrolman, and maintainer of trails.

The mountain may not have had a distinguished career in the lookout system, but it attracted sightseers before the Forest Service came into existence. A group of hikers gave Beljica its mysterious name in July 1897. They were five members of a nearby homesteading family—Burgon Mesler, his wife Elizabeth, and their three children, Isabel, Clara, and Alex.

With Lucy and Jessey LaWall, they climbed the unnamed mountain and acronymically christened it from the initials of Burgon, Elizabeth, Lucy, Jessey, Isabel, Clara, and Alex.

Logging roads have chewed away at the trail so it is just a mile to Lake Christine, where only a few rotten logs mark the former cabin site. The trail to the top is a bit beyond the lake.

Aerial view of High Rock Lookout

High Rock

Elevation 5,685 feet

High Rock earns its name by being the highest point on Sawtooth Range, just south of Mount Rainier. Although packtrains climbed a spine to the top—one animal at a time in the final few yards—the lookout building itself sits dramatically on the edge of a 1,500-foot drop-off to tiny Cora Lake.

Jim Hale spent three seasons on High Rock from 1938 through 1940. He made an interesting discovery there: petrified logs, found in a hole near the summit.

He liked the station except for one time-consuming chore that went with it. He had to pack water up to the lookout in a five-gallon can; and a round trip by the trail, on the gradually ascending back side of the mountain, took 2 to 3 hours.

Hale decided it would be faster to go up and down the sudden side. He ordered 300 feet of rope, bulky enough to require two mules to bring up the load. He uncoiled the rope through a chute and, using that aid, made it down to water in a matter of minutes. The ascent, hauling himself up hand over hand, took about 45 minutes.

When it came time to close for the winter, Hale coiled the rope and built a small shed over it. He was the only lookout in the Forest Service who had a collapsible trail, one that could be removed when he was done with it and restored in short order the following summer.

When the war broke out, Hale went into the Navy. While he was gone, two men stringing a new telephone line to the lookout chose the chute as a shortcut route. One fell and was killed. During the regional headquarters investigation that followed, the rope was discovered and dumped over the cliff for good.

Hale had a photograph of the lookout showing a protruding ledge on which he and others used to stand while looking down the drop-off. "That ledge is gone now," he said. "It just disappeared one day. Thinking about it gives you a spooky feeling."

Back in 1929 when the lookout was built, access to it was from Ashford over 10 miles of trail with a 4,000-foot climb. Even so, according to Hale, High Rock always drew a lot of visitors. Logging roads later crept up the valleys, and now the trail is a short 2-mile hike. Many more people now reach the still-standing lookout, including some who leave the door and shutters open to be wrenched loose by storms, and vandals who scribble obscenities.

High Rock is maintained by the Forest Service and manned each summer.

Aerial view of High Rock Lookout and Mount Rainier

Tatoosh

Elevation 6,310 feet

For a number of years, Tatoosh was the most famous lookout in western Washington. It got that way because of Martha Hardy's book, *Tatoosh*, published by Macmillan in 1946. The day-by-day account of a lookout's life was made more unusual because the firewatcher was a woman.

There had been a whole generation of women lookouts before her (she was on Tatoosh in 1943), and some of the others were also schoolteachers. They had been the subject of magazine and newspaper articles—but always written by someone else. Hardy's was a firsthand account, book-length, with a ranger district cast of characters, and it became a best seller. (It was reissued in 1980 by The Mountaineers.)

Hardy taught at a Seattle high school, but she owned a place near Packwood at the time.

The lookout, built in the 1920s and rebuilt in 1932, was on the highest

Cascades gold-mantled ground squirrel, possibly a descendant of Martha Hardy's friend, Impie

Tatoosh Lookout, 1939 (U.S. Forest Service photo)

point of the Tatoosh Range, just outside the boundary of Mount Rainier National Park. It was in sight of Backbone Ridge, Cowlitz Valley, Butter Creek Valley, and, to the west, High Rock. The town of Packwood is hidden by the lower part of the ridge, but the Cowlitz River and the highway along it can be glimpsed in spots.

The trail up is a killer. The sign says 5 miles, but add about 3,000 paces to that figure. The trail gains 3,000 feet of elevation in a long 3.5 miles, then traverses the range another 3.5 miles while gaining 1,500 more feet in elevation.

Lost Lake

Elevation 6,359 feet

Standing on the rounded summit of a ridge overlooking the forest of Coal Creek Valley, Lost Lake Lookout had views of Mount Rainier less than 20 miles to the northwest and the Goat Rocks peaks to the southwest. But the lake for which it was named was lost to view. Since the lake was there first, it could be argued that the lookout, not the lake, was lost. Today the lake is still there, but the lookout has been lost forever, removed in the 1960s.

While the standard hip-roof cabin was on a relatively flat area, the summit was surrounded by steep bluffs. After the building was erected in the early 1930s, the problem was how to get to it. Consequently, a trail was later blasted and hacked out of the cliffsides with pick and shovel.

Working on the Lost Lake Trail (U.S. Forest Service photo)

In 1944 the lookout was Martha Hardy, who had spent the previous year on Tatoosh. In her spare time she worked on the book she was writing about that experience. Two years later it became a best seller.

In 1935 Lost Lake became part of the Goat Rocks Primitive Area, which since 1964 has been designated the Goat Rocks Wilderness Area. It is extremely popular with hikers, and a large elk herd draws hunters. Hikers used to see big herds of the mountain goats for which the area was named, but when goat hunting was resumed in 1948 the animals became wary. What few are seen now are usually traveling alone.

Trails to the lookout are numerous—Coyote Trail from the White Pass Highway, Bluff Lake Trail from Purcell Creek, Three Peaks Trail from Coal Creek, and the Lost Lake Trail from Packwood. These were built for fire patrol, with the assumption that foresters could climb on steep grades quite as well as the resident goats. The lookout building was removed in the 1960s.

Burley Mountain

Elevation 5,310 feet

According to archaeologist Rick McClure, Burley Mountain Lookout was built in the summer of 1932 by William Mackay of the Randle Ranger Station. It is the last remaining hip-roof-style lookout still in Gifford Pinchot National Forest. The lookout is reached by miles of dusty road.

When other lookouts were being removed in the 1960s, Burley survived but was finally abandoned in 1974. In 1982, on its fiftieth anniversary, local volunteers, supported by Forest Service employees, repaired the building. Since then it has been staffed every summer by an employee or volunteer.

In 1989 the lookout became the summer home of Dan Ames, an opera

Burley Mountain Lookout and Mount Rainier

Sunrise from Burley Mountain

singer from New York. He is reported to have loved his mountain-top solitude where he practiced to his heart's content, entrancing the elk and eagles and especially the little ground squirrels who lived by his door.

The lookout is now rented to cross-country skiers and snowshoers in the winter. For information about volunteering or renting, contact the Randle Ranger District, P.O. Box 670, Randle, WA 98377 or call (360) 497-1100.

According to a 1989 letter from Ames, the lookout is an important radio relay system to the Gifford Pinchot Forest and Lewis County. It hosts over 800 visitors per season and is a popular launch site for hang-gliders.

Burley and Red Mountains are part of an air resource specialist program based out of Fort Collins, Colorado. Pictures are taken three times per day of Mount Rainier, Mount St. Helens, and the Goat Rocks, to determine air quality. With the use of a log book, data is recorded and information and pictures are shipped to Fort Collins. This has been in effect since 1987.

Burnt Peak

Elevation 4,135 feet

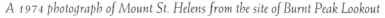

At the time the lookout was built in 1931, Burnt Peak (Burnt Mountain on some maps) was in the center of the largest de facto wilderness area in the state. It was roughly 25 miles by trail from the town of Cougar, due west on the Lewis River. The only road anywhere near the vicinity was a dirt track from Guler (a ranger station near Trout Lake) to the huckleberry fields in the mountains northwest of Trout Lake.

Burnt Peak does not rise high above the surrounding forest, and the lookout was not even on the highest point. The view to the north was limited, but to the east was Mount Adams, to the south was Mount Hood, and to the west was the Lewis River Valley and Mount St. Helens.

There are many equally good buttes and peaks for a lookout in this area, so chances are that Burnt Peak was chosen because it was crossed by the Cougar–Guler Trail. About 2,100 miles of trail threaded their way through the old Columbia National Forest (renamed Gifford Pinchot in 1949).

Burnt Peak remained the center of a wilderness until the 1950s. By then the loggers had finished cutting the productive lowland forest and moved to higher elevations. Two dams on the Lewis River created reservoirs, the upper one backing to within 8 miles of the mountain. Now it is surrounded by a network of roads, and the old trail system has been abandoned.

Away from the roads, traces of a short trail can be found among windfalls and young trees. The bare knoll where the lookout stood is reached in half a mile. The site is littered with melted glass, rusty nails, and steel cables. New growth is slowly encroaching on the view.

A 1974 photograph of Mount St. Helens from the site of Burnt Peak Lookout

Tongue Mountain

Elevation 4,838 feet

Juniper Ridge was the scene of a devastating fire in 1918. Thousands of acres were left with dead but standing trees, their thick trunks too green to burn in the passing flames. Over the years these dried out and became a ghostly "silver forest" rising high above the new tree cover below. In such cases, they become pitch-impregnated lightning rods during storms, and they often turn into torches. Tops or even the whole snag may topple and roll down the hillside, scattering fire.

In the mid-1930s, CCC crews cut many of the dead snags. As further protection for the returning forest, a lookout was built on Tongue Mountain in 1934. It was placed on an outcropping at the northern end of 11-mile-long Juniper Ridge.

The lookout was a short side trip from the Juniper Ridge Trail. The path

Tongue Mountain Trail and the Cispus River Valley

Tongue Mountain and Mount Rainier from Juniper Ridge

switchbacked up a very steep hillside to a saddle with an abrupt drop on the other side. The saddle must have been the end of the horse trail—a few old boards indicate that some kind of storage shed once stood there. A rough trail was hacked out of the last 100 feet of rock to the summit. The building was gone by 1948.

In mid-June Tongue Mountain is a large rock garden with bouquets of orange paintbrush, clumps of blue lupine, bobbing yellow wallflowers, various eriogonums, stonecrop, and bright rose cliff penstemon. There also are wild strawberries, white anemones, and a supporting cast of tiny white star-like flowers.

The trail is eroded in places by time and motorcycles. The final 100-foot climb is safe enough for an experienced hiker, but the ridge exposure is scary.

Mount St. Helens

Elevation 9,677 feet

One source says a Mount St. Helens lookout was planned for as early as 1915, but according to an old picture caption it was built in 1922, soon after the one at Mount Adams was finished. At 9,677 feet, it was the second highest in the state, and as might be expected at that altitude, weather proved to be fierce. St. Helens rose so far above its surroundings that it literally created its own weather.

People might be swimming and sunbathing at Spirit Lake beneath the mountain and admiring the cloud cap on top. To the lookout the "cloud cap" could mean a snowstorm and winds from 50 to 100 miles an hour, conditions under which he had no chance of spotting smoke. On other days, when it was drizzling at Spirit Lake, the lookout could be basking in the sun above a sea of clouds—which still gave him no chance of seeing a fire.

Getting lumber and supplies to the top was a major task. Scaling three of St. Helens' sides required mountaineering skills. Only on the south could horses be led up, and even that side had year-round snowfields interspersed with soft pumice. Old photographs show that at least some of the building

Opposite: Top, *A 1922 construction camp on summit of Mount St. Helens.* Middle, *Camping lumber on Forsyth Glacier* Bottom, *Lookout building* (U.S. Forest Service photos). Above, *Ruins of the lookout building in 1974*

materials were carried by men, up from Spirit Lake over the glaciers on the north side.

Eventually it had to be admitted that a lookout can get too high to be of much use. When the Forest Service gave up on St. Helens is not of record, but it probably was before the end of the 1920s. In 1937 the building had been long abandoned to snowdrifts, and by 1975 there was nothing left but a pile of lumber.

By the end of March 1980, the preliminary crater that formed at the summit prior to the May 18, 1980 eruption disposed of not only the debris but the site itself. It is safe to say that of all the lookouts ever abandoned, none other was removed so dramatically.

Many years back, conservationists had proposed a Mount St. Helens National Monument. Within days of the May explosion, an Eastern congressman came up with the same idea.

No doubt he envisioned building a new lookout from which visitors could see the full effects of the eruption. Therefore, in the interest of research for this book, we flew around the crater on the afternoon of July 22, 1980, looking for a likely spot on which to place an observation post.

The high south rim seemed best. From there, tourists could gaze down into the crater and out north along the path of destruction, especially on nice sunny days such as favored our flight. The lower part of the crater was a bit

Mount St. Helens and Spirit Lake, before the eruption

A minor eruption one month before the May 18, 1980 explosion

obscured by either steam or clouds, but otherwise the scene was peaceful—in an awesome sort of way.

Shortly after we landed back at Seattle, people were rushing to hilltops to watch the mushroom cloud rising down south. St. Helens had erupted again without the slightest warning, blowing steam and ash to more than 45,000 feet. We recommended that the crater observation post not be built there. The Forest Service must have seen the same cloud, because they built the visitor's center a safe distance away.

Silver Star Mountain

Elevation 4,390 feet

Silver Star sits a bit aloof from the rest of the Cascade Range, out west, overlooking the Columbia River, Camas, and Portland. North and east are forested mountains, but much of the timber is relatively new growth because Silver Star is in the site of the Yacolt burn—the most devastating fire that ever occurred in Washington. Aside from the immediate destruction of timber, thirty-five lives were lost and effects of the holocaust were felt for the next sixty years. Fires repeatedly struck through the burned area, making reforestation impossible.

The first burn in 1902 consumed 238,000 acres of prime forest, enough to build a three-bedroom house for every family then living in the state. A thick layer of smoke covered western Washington. In Seattle, 100 miles away, street lights were turned on at midday. In Olympia, at least one person panicked over "the eruption of Mount Rainier!" and was laughed at.

(The next time the midday sun was blocked out on such a wide scale, seventy-eight years later, "eruption" was no laughing matter. But it was St. Helens instead of Rainier.)

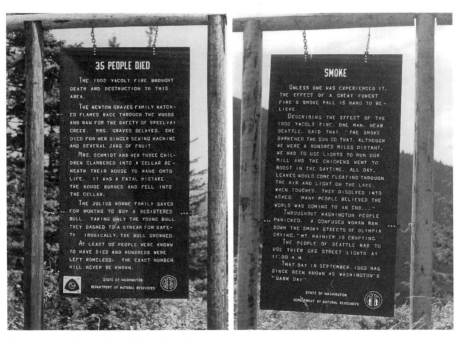

Signs along the road to Silver Star Mountain

Silver Star Mountain in 1978, with Mount St. Helens in the distance

The Yacolt burn led to the formation of the State Fire Wardens Association, which in turn became the State Department of Forestry, now called the Department of Natural Resources (DNR). It is only since 1965 that the DNR has been able to establish a new forest at the Yacolt site. To safeguard it, more than 600,000 snags were felled. Even so, a lot are left. They provide food and shelter for woodpeckers, but they also make targets for lightning.

An abandoned road climbs all the way to the lookout site, but one must walk the last mile or two. It is a pleasant walk, traversing lush alpine meadows with views of five volcanic peaks in Washington and Oregon.

Second lookout built on West Twin Butte (U.S. Forest Service photo)

West Twin Butte and Steamboat Mountain

Elevations 4,716 and 5,425 feet

East and West Twin Buttes are two rounded hills, perhaps ancient cinder cones. From a distance they are not distinctive but are isolated and high enough to view a large area of forest, and they were accessible from early roads. Both buttes were forested, so the top of West Twin Butte had to be cleared in 1923 for a 12-by-12 ground building with a cupola.

In the 1930s it was replaced by a standard hip-roof cabin, elevated on 12-foot legs. That one was burned in 1963 and trees are now covering up the view. But Adams, Rainier, St. Helens, and the rugged peaks of Sawtooth Mountain are still visible on the horizon. Below are the Mosquito Lakes, huckleberry fields, and miles of forest. West Twin Butte, along with Spencer Butte and Burnt Peak, was manned by teams working out of Twin Buttes Ranger Station.

Another nearby lookout was on 5,425-foot Steamboat Mountain. That building was erected in 1927 and when it was rebuilt in 1956, it was hoisted onto a high platform.

The tower timbers were 24 feet long, about the maximum anyone ever had packed up a switchback trail. Spencer Frey did it with a smart mule named Simon in the lead and a follower lugging the other end, an extreme example of an articulated mule team.

Bill Moran, a Forest Service employee, drew the assignment of burning down Steamboat in the mid-1960s. He recalled that there used to be a guard station a mile from West Twin Butte at Mosquito Lakes, and during the winter, equipment from various lookouts was stored there. Stacks of boxes were marked "Burnt Peak," "Point 3434," "Twin Buttes," "Steamboat Mountain," and "Switchback."

Puppy (Dog Mountain)

Approximate elevation 2,400 feet

In its lengthy journey south along the backbone of the Cascade Range and into the Sierra Nevada, the Pacific Crest Trail goes right over the top of some peaks that stand on ridges. It dips again and again into gaps and passes that, even as low spots, are a mile or more in elevation. It crosses Washington's five highway passes at their summits, dropping to 3,000 feet at the lowest one, Snoqualmie.

Then the Crest Trail arrives at what is perhaps the world's most extraordinary mountain pass, the Columbia River Gorge. There are many rivers bigger than the Columbia, but they do not cut their way directly through a major mountain range. Here, in the good old days, Crest Trail hikers went back to square one, descending to 186 feet above sea level and starting over again on the other side of the river.

Until 1985 the Crest Trail went over the top of Dog Mountain where the trail began to drop into the huge "pass." The lookout was placed on a spot shoveled flat along the ridge leading to 3,100-foot Dog Mountain. It had a sweeping view of the Columbia, 2,000 feet below, and of forested hillsides over in Oregon. Since the gorge is as deep as Dog Mountain is tall, the only peak in sight at the lookout point is the tip of Mount Hood.

The Crest Trail now starts near the Bridge of the Gods. The old trail to Dog Mountain can be picked up from the Columbia Gorge Highway, US 830, between mileposts 53 and 54, and followed 3 miles to the lookout site. The lookout was called "Puppy" because it was only part way up a rather small (3,100 feet) mountain, and a Dog Mountain at that.

Columbia River from the site of Puppy Lookout

Sleeping Beauty

Elevation 4,907 feet

Sleeping Beauty, on a southwest ridge of Mount Adams, stands out on the skyline when seen from Trout Lake, and it was from that perspective the mountain got its name. Someone visualized a woman sleeping peacefully on her back, but the romance inherent in the notion was somewhat diminished in 1931 when the Forest Service built a lookout right on her nose.

From then on, as one approached the point where the lookout stood, it appeared that a more appropriate label might be The Monastery. The building was perched on a narrow rocky ridge with 200- to 400-foot drop-offs on all sides. The trail up the cliffs is a wide path built on handmade rock walls, some more than 10 feet high. The rocks were carefully fitted together without mortar. Surely an order of monks, or maybe followers of a high priest who lived on the peak, did the work.

It is said that at one time the building was surrounded by a fence, which would have been a reasonable precaution against taking an extra step away from the house. The building is gone but the cable anchors that held it are still there, which indicates that the Forest Service constructed it after all. Ancient monks wouldn't have had steel cable.

Left, *Sleeping Beauty.* Right, *Rock bulkheads needed to support the trail to the lookout*

Sleeping Beauty Lookout (U.S. Forest Service photo)

As further proof that the Forest Service did it, the Mount Adams Wilderness Institute (operated by the Lloyd family of Glenwood) has on record an article written in 1931 by E. G. Hayes. He said it took 11 days for packtrains to carry supplies from a Glenwood mill to the 4,200-foot level, where a camp was set up.

The materials were winched the final 350 feet by a hand-operated windlass and 700 feet of cable. According to Hayes, the cable broke under the strain of one load and thirty pieces of lumber were reduced to kindling by the fall.

Ernie Childs, a Forest Service packer at the time, says the building was pre-cut and packaged by a Portland firm. He transported the bundles to the base camp and helped winch them to the top. Tony Guler, Louie Jarvis, and Don Williams assembled the building and Williams became the first lookout.

The once long journey to Sleeping Beauty has been shortened to an arduous day's walk by logging road. The view is much the same: miles of timberland from the slopes of Mount Adams to Sawtooth Mountain and down the White Salmon River to the farms near Trout Lake. Mount Hood and Mount St. Helens add to the landscape.

Summit Prairie and Dark Mountain Trail

Elevations 5,238 feet and approximately 4,900 feet

Summit Prairie Creek spills off Quartz Creek Ridge, but it must have been named for some point on the ridge other than the highest one. That wooded knoll was hardly a prairie, at least not until several hundred trees were cut to give the site a view when the lookout was built in 1929 or 1930.

A quarter of a mile away was another lookout of sorts, Dark Cabin, a cedar shake house on the Dark Mountain Trail. (The USGS map called it Summit Guard Station.) The cabin had a fire lookout platform on its roof.

Dark Mountain and the adjacent Dark Meadow were named for John Dark. During winters he and William McCoy worked a gold mine in the Lewis River drainage. That accounts for McCoy Peak and McCoy Creek.

McCoy's sons packed the older men in and brought them out in the spring. An indication of how deep the snow got was the height of stumps of trees the miners had cut for firewood. Some were 15 feet tall. Keith McCoy of Trout Lake, the third generation, was the area's unofficial historian.

Views at Dark Cabin were somewhat restricted, but Summit Prairie looks out on the forests of the Quartz Creek Valley, the upper Lewis River, Juniper Ridge, and those south to Sawtooth Mountain. North, south, east, and west, almost at the four compass points, stand three giant volcanoes—Rainier, Hood,

Mount Adams and the remains of Summit Prairie Lookout

Top, *Summit Prairie Lookout.* Bottom, *Dark Cabin. Note the ladder and observation platform on the roof.* (Photos by Dick Moran)

and Adams—and one former giant, St. Helens, who is 1,300 feet lower than she used to be because she now goes topless.

Another feature on the horizon was the subject of a notice posted on the firefinder pedestal: "In case of emergency, point finder at azimuth 84 degrees, 25 minutes, vertical angle 9 minus." When forester Bill Moran followed the instructions, he found the finder pointed to the outhouse door handle.

The lookout was manned until 1963 and used by trail crews until, along with Dark Cabin, it was removed in 1967. Concrete foundation blocks stand as a memorial, a row of solemn cubes that look like Druid columns just beginning to sprout.

Madcat Meadow

Approximate elevation 5,800 feet

Madcat was a patrol station on a southern shoulder of Mount Adams rather than an actual lookout. It consisted of a storage shed and a wooden frame on a rocky outcropping to put the firefinder on. Southward it scanned the White Salmon River Valley forests down to Trout Lake, and to the west Sawtooth and Steamboat Mountains. The view to the north and east is blocked by giant Mount Adams and the ice cliffs of Avalanche and White Salmon Glaciers.

Madcat began about 1927 when the patrolman at Gotchen Creek Guard Station rode horseback around the territory to check up, among other things, on the sheep bands in the Mount Adams area. There were many of them, and according to Keith McCoy, they trod out Round-the-Mountain Trail, which was also the original route of the Forest Service telephone line from Trout Lake to Randle.

Shelters similar to Madcat existed at Divide Meadows and on the southern slope of Adams at Snowplow Mountain. Their materials were packed in by Julius "Jude" Wang, who rented the horses from Wyers Stage Company of White Salmon. By then stage companies were running on a different kind of horsepower, but apparently Wyers still kept the old type as spare parts stock.

A few bleached boards are all that remain of the Madcat firefinder stand, but the storage shed is still there, badly bent by heavy snows.

Storage shed crushed by snow at Madcat Meadow, 1978

Flattop Lookout and Mount Adams (U.S. Forest Service photo)

Flattop Mountain

Elevation 4,394 feet

This accurately named mountain overlooks the town of Trout Lake and the farmland around it, with Sleeping Beauty and Mount Adams to the north and Mount Hood down south.

The lookout is the second one to occupy the spot. Well designed, with tinted windows sloping outward to reduce reflection, it represented the two-story model and (rather ahead of its time) the flat-roof style. Due to land swaps, the building now stands on state land. It is abandoned and deteriorating and trees are slowly encroaching on the views.

A short drive across the timbered flat top brings one to a view, on the west side, of more forest land and Mount St. Helens. This site had been used as an observation point, with a firefinder on location so the lookout could walk a quarter-mile and sight around the western area. When the original eastside building was replaced by the present structure in 1935, it was moved across the flat to this westside location. It was one of the last buildings to go when surplus lookouts were destroyed. The first human lookout, in 1922, was Amos Pearce, from nearby White Salmon.

Ernie Childs helped construct the first building. He said the materials were hauled up on wooden sledges from Deadhorse Meadows, west of the summit, by Aubert Robbins who owned a sturdy team of bays. The route was the same as today's road to the top.

It is an old road, roughed out soon after the first lookout was built. It shows on a 1924 map updated in 1930. The final spur to the top starts off as an unmarked but well-graded logging road, then falls back into its original condition, a rough, steep dirt road.

Mountaineer C.E. Rusk on the summit of Mount Adams in August 1921 when the lookout was under construction (Photo by Joseph Vincent from the Whitnall-Millen Collection)

Mount Adams

Elevation 12,276 feet

Back in Washington, D.C., when the lookout system of the future was being plotted on the drawing boards, it was noticed that the Cascade Range offered a great advantage—peaks that towered so tall, anyone on top should be able to watch thousands of square miles of forest.

Mount Adams, for example, was 12,276 feet, higher than airplanes could fly. A lookout was planned for it and for other dandy volcanoes. In 1913 they started with Mount Lassen in California. It erupted in May 1914, and bounding boulders busted the building.

Undiscouraged, Forest Service headquarters put Mount Hood on the schedule for 1915, along with another serene volcanic peak, Mount St. Helens, but work actually began at Adams in 1918.

From Trout Lake, the summit was reached on the relatively gentle south side of the mountain, across soft pumice fields and permanent snowfields.

Although horses could be used partway, the supplies then were winched to the top, a few hundred feet in each lap. It took another three years to do that and finish the building.

Later on, miners built a trail, parts of which can still be found. They were interested in the sulphur deposits which, near the top of the south rim, are 10 to 15 feet deep and 90 percent pure. The miners figured there was well over half a million tons there.

A couple of difficulties faced the miners. One was water for the drills—they had to melt the snow—and the other was getting the stuff down from the top. That final step was never successfully accomplished, although sporadic activity was continued until 1959.

Since Adams is farther east, the low cloud cover that plagued St. Helens and Anvil Rock, Mount Rainier's highest lookout, didn't occur quite so often. Nevertheless the peak worked up storms at any time of year, reducing visibility to zero. Attempts to use the highest-of-all lookout lasted no longer than it had taken to build it. It was not manned after 1924.

Earl Dean of White Salmon—whose father, Wade Dean, had staked claims

This photo taken August 1, 1937 shows the completed lookout building and the addition built by miners. (Photo by Donald G. Onthank)

In September, 1981, after a mild winter and hot summer, mountain climbers discovered the partially exposed remains of the lookout building. (Photo by Vicky Spring)

over the entire summit by 1929—built a lean-to (long gone) in 1932 when work began on the claims. In September 1936, he was pinned down there in a storm for 10 days. In the summers of 1935 and 1936 the miners used the abandoned lookout, which was filled with snow and ice each winter.

Since the mid-1940s the inside of the building has been packed solid with ice and snow, and the entire building is buried under a snow drift, which probably accounts for how well the building is preserved after all these years. Most years, by late summer the snow has melted and part of the roof is visible. A few times the building itself has become partially exposed.

There are many climbers each year but they don't arrive in a huge herd, as they did each summer from 1966 to 1976 when the Yakima Chamber of Commerce decided to turn Adams into its own Fuji and conducted an annual march to the summit. On at least one ascent, there were more than 500 participants. After 1976, the Forest Service put a limit on the number in an "assault party." Since then the "Pah-To" (as the Indians knew it) has let Fujiyama go unchallenged in the pilgrimage records.

5

WENATCHEE NATIONAL FOREST

One hot afternoon the lookout on Sugarloaf, a 5,814-foot peak on the boundary between the Leavenworth and Entiat Districts, spotted smoke rising from the Chumstick Valley. He reported it to the closer district's headquarters, Leavenworth.

After a few minutes Leavenworth called back. The smoke also had been reported from Beehive, some 25 air miles to the southeast, but with a long view of the Wenatchee Valley. Beehive had the fire in a different location. Would Sugarloaf, much nearer, take a second siting?

The Sugarloaf lookout went back to his firefinder. At first he couldn't see

Poe Mountain Lookout (U.S. Forest Service photo)

Top left, *Domke Mountain Lookout*. Top right, *Thorp Mountain Lookout*. Bottom left, *Goat Peak Lookout, near Cle Elum*. Bottom right, *Junior Point Lookout* (U.S. Forest Service photos)

the smoke, then it suddenly showed up again, right where he had placed it before. He confirmed his first reading.

Again a silence from Leavenworth, then came the word that the man on Beehive also had taken a second look and insisted his initial report was right on the button. Both lookouts continued to plot the location and both remained adamant about their accuracy. They agreed only that the smoke was intermittent, but when it rose above the trees it really boiled up.

The ranger decided to send a truck up the Chumstick Road to find out what was going on. It was going on forward, then backing up. When the coal-burning locomotive, working with a roadbed crew, came out one end of a tunnel, Sugarloaf was right. At the other end, Beehive had the smoke pinpointed.

That lookouts so distant from one another could zero in on the same smoke was due to the topographical layout of the Wenatchee National Forest. Its major valleys start in the high Cascades and parallel one another on a northwest-to-southeast diagonal.

The forest is bounded on the north side by Sawtooth Ridge, with peaks up to 8,500 feet paralleling Lake Chelan. South of the lake is a 50-mile string of peaks called the Chelan Mountains. Next come the Entiats, then the lengthy Wenatchee Mountains that stagger southeast from the Cascade Crest to the Columbia River. Separating the Yakima and Naches River Valleys is the Manastash Ridge, with the final region of Tieton running west to east from Mount Rainier National Park toward the city of Yakima.

The seven ranger districts of Wenatchee are laid out in pretty much the same way, each taking in a major drainage area. Technically the Naches and the Tieton Districts are in the Mount Baker–Snoqualmie National Forest, but since it requires an act of Congress to change official boundaries, the Forest Service takes an easier way out. It falls back on common sense, with the Wenatchee "administering" these territories that, being east of the Cascades, obviously belong in its domain.

Wenatchee National Forest lookouts claim a number of "records" of one sort or another. Perhaps they can be challenged, but let the facts speak for themselves:

Badger Mountain was manned from 1953 to 1974 by Oscar Richardson of Cashmere, who started his Forest Service career on Beehive in 1952. His twenty-two years at the same station may be matched by others—but who else became a lookout at fifty-four and stuck to it until he was seventy-seven years old?

At the other end of the scale, time once appeared to stand still at the Dirtyface Lookout, near the upper end of Lake Wenatchee. Jimmy Burgess, a big youth who got the lookout job there in 1947, said he was sixteen—the minimum hiring age. Next year when he signed up for the same post, he put down his age as sixteen. By the summer of 1949 the district ranger began to wonder if the spring from which the lookout drew his water was the fabled

Icicle Ridge Lookout

Fountain of Youth. For the third year Jimmy was sixteen—and for the first time, honest about it.

"He was a sharp kid," says Jim Currie, who moved from Estes Butte to David to Sugarloaf during those years. "To avoid any questions about his age or competence, he tried to beat everyone else at spotting fires even when he was supposed to be taking a rest break, and often he succeeded. When we finally found out how old he was, we forgave him."

Tiptop, near Blewett Pass, could be renamed "Kester Mountain." After Harold V. Kester spent seventeen years there, his son Estes took over. As far as we could find, no other district can boast a consecutive two-generation claim on one mountain.

The first aerial supply drop was made on the Rat Lakes fire in 1939, according to Ken Wilson of Leavenworth, the fire boss at the time. The drop was done by a Ford tri-motor, and Wilson remembers Kenny Patton to be the pilot. Burlap squares were rigged by rope to the four corners of supply boxes to form parachutes, which worked with varying degrees of success. Eggs were purposely scrambled before delivery, according to Wilson, though the author recalls a similar drop in the Snoqualmie National Forest around the same

McCue Ridge Lookout

time. The eggs there were packed in the middle of loaves of bread, and survived unbroken. (Incidentally, the Rat Lakes were later renamed the Enchantments, to give them a more appealing sound.)

Another aerial distinction, although not of Forest Service origin, used to delight passengers on the morning flights from Seattle to Spokane. In those days (the 1930s), the DC-3s barely cleared the Cascades at Snoqualmie Pass, and the pilot would pitch the morning newspaper out the window to the lookout at Snoqualmie. That lookout became the most up-to-date man in his district.

It is doubtful that the first woman lookout ever can be named for sure, but again Wenatchee National Forest claims precedence. Possibly more than one started in the same summer, 1917, when there was a shortage of young men because of the First World War. But by the time they began appearing on the job in other districts, women lookouts were considered no great novelty in Wenatchee.

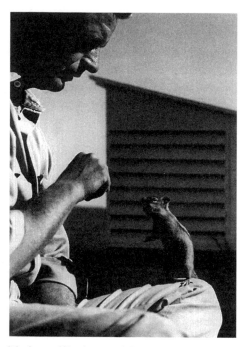

Timberwolf lookout and friend

Iva West (later Gruenewald) spent the summers of 1920, 1921, and 1922 in a tent near the top of Tumwater, which looks down on Tumwater canyon and Leavenworth. She said she was on the peak above the tent site from 8:00 A.M. to 5:00 P.M. daily, and during the first summer reported about ninety fires. Many of them had been started in the grass along the Wenatchee River after a Great Northern train had passed by. Other members of Mrs. Gruenewald's family followed her lead. Her sister, Gladys West, spent a summer at Sugarloaf. A brother, Talbot West, took over Tumwater when Iva left it, and later spent summers on Dirtyface and Sugarloaf.

During that period the forest supervisor was A. H. Sylvester, who hung more names on the national forest map than any other man in history. He was supervisor from 1908 until he retired in 1931; he died in 1944 from injuries received in a trail accident.

When he began, much of his domain was not mapped in detail, leaving to him the job of labeling lakes, streams, ridges, and peaks as he came across them while surveying the territory. He was something of a scholar, which accounts for the string of poetic peaks—Bryant, Longfellow, Poe, and Irving, as well as the Biblical set of David, Saul, and Jonathan.

The contour lines of one unnamed mountain were so close together they reminded him of a maze. He thought of Theseus entering the labyrinth to kill the Minotaur, so the state has a Labyrinth Mountain with Lakes Theseus and Minotaur on it.

On a 1909 trip, Sylvester was accompanied by Ranger Burne Canby, who wanted to name three lakes they discovered for his sisters, Margaret and Mary, and a friend of theirs, Florence. Sylvester said that was okay, if the next lake they found would be Alice, after his wife. They found it.

It set a precedent. As Forest Service personnel came across lakes without a name, wives, mothers, girlfriends, sisters-in-law, and cousins were entered on the map. Wenatchee National Forest ended up with more than a hundred female names attached to its topographical features.

Davis Peak

Elevation 6,426 feet

Davis Peak is a high point on the south end of a lengthy ridge called Goat Mountain. It looks along the Cle Elum River Valley from Salmon la Sac north, and to the northwest up the Waptus River to Dutch Miller Gap and the Cascades. To the east, Mount Stuart rises to 9,415 feet.

Grover Burch, one of the first rangers in the Forest Service (he joined in 1908), became the Cle Elum District ranger in 1926. He named the peak for Louie Davis, a young lookout who had died during the off-season. The station was built about 1934, and in 1944 it was occupied by Marge Lumsden, a schoolteacher and fourth-generation resident of the Roslyn area.

She expected no career as a lookout; she was just filling in until the men got back from the war. They did—but four years later came Korea. Marge worked in the Wind River District for one summer, then returned to Davis Peak for the rest of its active life, putting in a total of sixteen years there.

The building was surrounded by rocky open spaces and heather, and Marge used to entertain herself watching the mountain goats that grazed around the "yard."

In 1965 there was so much snow she had to be landed on Davis by helicopter. That was the year the Forest Service decided to abandon most lookouts, and it proved to be her last season. The building on Davis Peak was removed in 1968, but the trail to the lookout site is unchanged from the time Marge first used it.

Mountain goats near Davis Peak Lookout in 1958 (Photo by Marge Lumsden)

McGregor Mountain

Elevation 8,122 feet

Even with a bit of its top blasted off, McGregor Mountain rises more than 8,000 feet, high above the Stehekin River Valley. When Ellis Ogilvie was working for the Forest Service in the early 1940s, he overheard bits of radio conversation between McGregor Mountain and the Stehekin Ranger Station. The lookout talked about mountain goats, glaciers, and glimpses of the headwaters of Lake Chelan.

It sounded like a great place. The next season, 1944, Ellis applied for the position. His supervisor stared at him silently before observing, "Well, I once asked for that job. That makes you the second one ever to volunteer for McGregor."

Packing in took two days. At the trailhead there was 6,500 feet of elevation still to gain. The climb was over 127 switchbacks in the first 7 miles, and nowhere was water nearby. At 7,000 feet Ellis and the packer reached a small meadow, beyond which the peak rose precipitously for another 1,100 feet.

"This is as far as I go," the packer announced. Pointing at a 700-foot cliff, he added, "There's the trail." He began unloading Ellis's supplies, which included 100 gallons of stove oil to keep him from freezing up there.

Ellis stared at the cliff. "What trail?" he asked.

"See those red paint splashes? They mark the best ledges to climb on."

According to Bill Lester of Winthrop, who helped put the cabin with cupola on the peak, it was done with cables to tram up the materials. The lookout was sure to be a hermit for the season, but that wasn't all. During lightning storms, Ellis recalls, "the network of copper cables around the cabin carried so much St. Elmo's fire, it looked like a Standard Oil station lit up with blue neon tubes."

This Siberia of lookouts had an intermittent record of occupancy. A *National Geographic* magazine article in the early 1950s showed a packtrain on the rock-jumbled sides of McGregor. Forest Service men were packing in a college teacher who had decided she wanted to be a lookout and write about it. She later stuck with that kind of job for three summers, but on McGregor she held out for less than two weeks.

The mountain is just inside the boundary of North Cascades National Park. When the Park Service took over and burned the lookout, there were few if any mourners.

Yet the peak continues to be used for another purpose. A radio repeater station was built, with the help of a helicopter instead of packers. Nevertheless, the helicopter pilots who brought up a serviceman twice a year joined the packers' brotherhood after they nearly crashed. They refused to try again, and landed on the ridge below. Now the servicemen climb the last 1,100 feet, like the lookouts before them.

With the time it takes to get to Stehekin and on up the valley to the trailhead, and the long climb over the thirsty switchbacks, it remains an overnight trip even to the packer-and-helicopter pad at 7,000 feet. The red paint splashes have eroded away too, so the final mile to the top is, if anything, a tougher climb than it used to be.

McGregor Mountain Lookout, with its antennae-like lightning rods

Taking a panorama photograph from the roof of Pyramid Mountain Lookout (Photo by Goodman)

Pyramid Mountain

Elevation 8,247 feet

The view from Pyramid Mountain is range after range of snowcapped mountains in the Glacier Peak Wilderness, a near-vertical view of Lake Chelan more than 7,000 feet below, and the long sweep of Sawtooth Ridge beyond that. The mountain stands midway along the lake on the south side. At 8,247 feet, it is still junior to five taller neighbors along the Chelan Range.

Logging and service roads have obliterated all but a small portion of the former trail system that switchbacked up the barren western half of Pyramid, a rounded hill. The east side is a nearly straight drop-off. All that is left at the lookout site is a stone foundation, a small storage shed made from rocks, a hitching rail for packhorses, and a houseless outhouse. The seat remains in position and is still usable.

An undated photograph at Pyramid Mountain shows six packhorses laden with lumber, led by a saddlehorse and rider. A second, unmanned saddlehorse must have been the photographer's mount. When he took that scene, it was incidental to his primary assignment. In a later photo he is shown teetering on the roof of the building with his huge circuit camera.

This type of camera was loaded with film lengthy enough to record 360 degrees, a portion at a time. The operator set it up and triggered the shutter; then the camera revolved automatically, snapping at each halt around the compass. An advantage over a wide angle lens was that it did not distort the picture.

Topographic features could be labeled on panorama photographs and both headquarters and the lookout had copies. They were a quick reference to what could be seen from up there.

Many such pictures were taken to help determine where lookouts should be placed. In that case the photographer had to lug his heavy equipment to places and up peaks where no trail had yet been built. His job was no snap.

Packtrain headed for Pyramid Mountain in the 1930s (Photo by a government packer, courtesy of his son, Dick Goodman, a U.S. Forest Service employee)

Sugarloaf Peak

Elevation 5,814 feet

The lookout site at Sugarloaf, in the Entiat Mountains north-northeast of Leavenworth, may be the one in longest continuous use in Washington State. Lookouts lived in a tent before the first building was erected almost ninety years ago; and today Sugarloaf is a rarity—a station that still stands and is still active.

The peak is a cornerpost for three ranger districts—Lake Wenatchee, Entiat, and Leavenworth. Forests stretch for miles in every direction—although not quite as far as the eye can reach, since the wheat fields over across the Columbia River are in sight. The end of Lake Wenatchee is in view 10 miles to the west, and high mountains in the distance include Mount Stuart and Glacier Peak.

Until the early 1920s, Sugarloaf was a remote outpost. Then a road to it was roughed out when the first building, a 12-by-12-foot cupola type, was put up in 1924. In the process a tall rock pillar on which the firefinder previously had been mounted was blasted away to make room for the building.

Over the years the road was improved, and once sightseers could drive there for the panoramic view, they did. By the late 1940s they arrived in

During lightning storms, the firewatcher stands on an insulated stool.

Sugarloaf Peak Lookout, 1978

Sunday parades of up to fifty or sixty persons. Public attention turned Sugarloaf into a Forest Service showcase, and the lookout had to behave accordingly. Saturday and Sunday schedules called for being up and dressed in uniform, necktie and all, by 6:00 A.M.

When the Lion Rock Lookout near Liberty was closed, its cabin—a hip-roofed 14-by-14—was sawed into sections, transported to Sugarloaf, and re-erected in place on the old cupola building. Thus a claim can be made that the present lookout structure has watched over more square miles of land than any other Forest Service station. It was moved far enough that its views never overlapped, so during its two lives it has looked all the way from Ellensburg on the south to the mountains on the hazy horizon north of Lake Roosevelt in the Colville National Forest.

Sugarloaf is on a Forest Service road that snakes along or close to the ridge of the Entiat Mountains for 25 or 30 miles. Chumstick Mountain, another busy lookout of the past, is 10 or so miles of winding road to the south.

155

Poe Mountain

Elevation 6,007 feet

P oe Mountain is one of the high points on Wenatchee Ridge unofficially known as "Poets Ridge" because several other peaks on it are called Bryant, Longfellow, Irving, and Whittier. The lookout at Poe had a commanding view of the Little Wenatchee River Valley from Meander Meadows to Soda Springs, the forest on Nason Ridge, and the mountains along the Cascade crest. Views in other directions are partially obscured by various poets, but Glacier Peak, Sloan Peak, Monte Cristo, Mount Hinman, and Mount Rainier can be seen above distant ridges.

When seventeen-year-old Jim Currie drew Poe Mountain in 1946, it was his second year as a lookout. He hadn't smoked until then but he felt his veteran status called for two cartons of cigarettes among his supplies. H. E. Peters, the district ranger, spotted them during an inspection trip. He said nothing then, but after the ranger departed from a second visit, Currie found the cartons were gone. Peters was dead set against cigarette smoking in his district.

"I haven't smoked ten cigarettes since then," Currie commented. But before the American Cancer Society plants a plaque on Poe to commemorate the event, it should know that Ranger Peters was a pipe smoker. In place of the cigarettes he left a pipe and a pound of tobacco.

Poe can be reached from the end of the Little Wenatchee River road on a trail that climbs 3,000 feet, or from Forest Service Road 2817 with about a 2,000-foot climb. Either way it is not more than a 3-mile hike but it is a dry passage, all exposed to the sun, and those who had business at the old lookout preferred to hike the distance after sunset. The building, gone except for some bits of rubble, was a standard R-6 from the early 1930s.

Wenatchee Ridge and Glacier Peak from Poe Mountain

Icicle Ridge Lookout (U.S. Forest Service photo)

Icicle Ridge

Elevation 7,029 feet

The lookout, built in 1929, was situated on 12-mile-long Icicle Ridge, which divides the Icicle River and Chiwaukum Creek drainages. Southwest across the 4,000-foot-deep Icicle Valley is 8,500-foot Cashmere Mountain. On to the south is the dramatic Stuart Range and the glaciers of 9,415-foot Mount Stuart itself.

The high spot on the ridge, a sharp outcrop pointed at the sky, offered an inhospitable building lot. Dynamite chewed it down a dozen feet but its new flat top still provided only a sparse 144 square feet on which to set the cupola-style lookout. There was no side yard and access to the cabin door was by ladder. Ken Wilson, who manned the lookout in 1936 and 1937, encircled the cabin with a catwalk and replaced the ladder with stairs.

In 1963 the Forest Service talked about rebuilding the lookout in another location, but decided in favor of a patrol instead. The building was abandoned in 1964 and was burned about 1969. The debris marking the spot includes half-burned boards, pieces of batteries, iron, and loads of nails scattered about.

The site is commonly reached by the steep Fourth of July Trail from the Icicle River. A sign says the distance is 5 miles, but it could easily be 7 or 8 miles, in a climb of 4,600 feet.

Mount David Lookout, 1970

Mount David

Elevation 7,420 feet

The White River starts at the crest of the Cascades near the southern end of the Glacier Peak Wilderness Area. As it flows southeast, it picks up tributaries from a 150-square-mile basin. After 30 miles it pours into Lake Wenatchee, and there it suffers the fate of many rivers that seem to be headed for major status: the outlet of the lake becomes the Wenatchee River.

Mount David is the highest peak near the center of the White River basin, so it was chosen for a lookout station in spite of the difficulty in getting to it. It is a 5,200-foot climb to the lookout, and 3.5 miles of incessant switchbacks lead only to the ridge, not even the halfway point in the hike. The trail then goes up and down the ridge, sometimes on narrow ledges blasted out of the cliffs, with snow and ice until August.

Packtrains had to stop 2 miles short of the peak. Steel handrails were installed to help lookouts pack up from there, but the rails did not long survive avalanches. They were gone by the time Jim Currie was on Mount David in 1948. "Whoever got that summer assignment was the 'aristocrat' of the lookouts," says Currie. "He could boast he had the meanest station in the district."

The cabin's "yard" was 2 feet wide, penned by a cable. The lookout could drop a rock out the west window and never hear it land 1,800 feet below. The eastside drop-off was gentler, not more than 600 feet. But the view, which included the top of 10,541-foot Glacier Peak 12 air miles to the northwest, was great.

The cabin, built in 1934, was abandoned in the 1960s and burned down sometime after 1972. Forest Service helicopters removed the debris. Lookout Magnus Bakke gets credit as the outhouse architect. It was built in the 1930s with a hole 500 feet deep, and still remains.

Red Mountain

Elevation 5,707 feet

From Keechelus Lake on the west to the Sasse Ridge summit on the east, the Cle Elum Ranger District had as complete a set of lookouts as could be found anywhere. Between Kachess and Cle Elum Lakes the posts were only 2 or 3 air miles apart, so there were few crannies among the lakes and over to the Cascade Crest that could not be put under surveillance.

Red Mountain, north of Cle Elum Lake, was in one of the chummiest spots. From its summit, the person there could see five other lookouts—on Thorp, North Peak, Davis, Jolly Mountain, and, if he climbed higher, Pollalie Ridge. The building itself was not at the 5,903-foot top of the mountain. It was a mile south and east, on a grassy knoll at the 5,707-foot level, perched on the edge of a 700-foot drop-off into Thorp Creek.

The site had been used during emergency periods for some years before a building was set up, as evidenced by a Forest Service panoramic photo taken from there in 1934. But the fate of the Red Mountain Lookout was prophetic of things to come. The standard R-6 building was little more than ten years old when it was removed in 1948, on the grounds that it wasn't worth its maintenance cost. It had not kept up with its neighbors in initial sightings.

Joe Ostliff, who had packed in material to build many of the lookouts, packed most of this one out again. It was dismantled and even the windows were saved. Joe explained why there is no trace left of what little remained: "We threw gasoline and a match on it and ran like hell."

The trail once started from the Salmon la Sac Guard Station but has been shortened a few miles. It is very steep and near the top becomes quite obscure.

The view from Red Mountain: Cooper Lake and the mountains along the Cascade Crest

Badger Mountain

Elevation 3,498 feet

Only one lonesome fir stands anywhere near the lookout. Surrounded by farmland and sagebrush on the east side of the Columbia River, the building sat on a low, stocky tower. At night you can see the city of Wenatchee to the northwest sparkling with lights, and cars threading their way down US Highway 2 from Cashmere. This is a Forest Service lookout? It isn't even in a national forest! However, it has a sweeping view of the lower ends of two districts across the river, Entiat and Leavenworth, from Mission Ridge almost to Chelan. The Badger Mountain Lookout has rung up its fair share of initial fire reports.

Behind the tower, and slightly higher, is a weatherbeaten fir with its top sawed off. This was the original lookout, and part of the ladder is still nailed to it. The lookout is surrounded by private land, whose owner finally tired of picking up beer cans and repairing fences and has gated the road.

Oscar Richardson all but monopolized the history of Badger, at least after the building replaced the tree as the lookout station. He served twenty-two years there, up to 1974. He spotted fires as far away as 35 miles.

A firm believer in an ancient signaling device, the heliograph, he once

Badger Mountain Lookout and a sailplane carried on updrafts along the ridge top

Fir tree originally used as the lookout, with the town of Cashmere in the distance

directed a fire crew 32 miles away by flashing his mirrors to the source of the fire. One time he got his signals through to the Dirtyface Peak Lookout above Lake Wenatchee, 40 or more miles distant.

That range of visibility should explain both the seemingly oddball location of the Badger Lookout and why it was manned as late as 1986.

One of Richardson's shorter-term predecessors was Charlotte Raine from the ranch directly below the station. Then came Barbara Calde, who watched from Badger during 1949 and 1950. At the end of the second season she met a young fellow who had just closed his lookout on Sugarloaf. She and James Currie became one of many such pair-offs in the Forest Service.

Badger Mountain sits crossways to the prevailing west wind and its dependable updraft is popular with both birds and gliders. Glider pilots ride for miles along the ridge, and hawks zigzag above it for hours looking for mice. Other birds hover motionless, seemingly for the fun of it.

The only wildlife Barbara Currie doesn't particularly care to remember were the rattlesnakes in the woodpile.

After being unused for ten years, the lookout was trucked to Entiat to become part of the Columbia Breaks Fire Interpretive Center.

Thorp Mountain Lookout

Thorp Mountain

Elevation 5,854 feet

Thorp Mountain Lookout sits on Kachess Ridge, which runs between Kachess and Cle Elum Lakes. On the Cascade side the views include the Kachess reservoir 3,600 feet below, Mount Rainier, and the Dutch Miller Gap mountains. On the eastern curve, tiny but picturesque Thorp Lake adds to the scene, and Mount Stuart is part of the horizon.

Thorp is topped by extensive alpine flower fields, all in bloom at the end of the snowmelt which usually comes by the end of June. During the summers of 1945 and 1946 a schoolteacher, Bea Buzzetti, was the lookout. She made further use of her time to gather, identify, and dry ninety-two varieties of plants. She presented the collection to the Forest Service, and the Thorp Mountain Herbarium still can be seen at the Cle Elum Ranger Station.

Rather than pack water from a mile away, Bea diligently melted snow each day. Finally, she could gloat over the forty gallons she had stored in a canvas bag hung in a tree. Then, when the snow was gone, her 320-pound treasure broke its rope and splashed away. After that she could gather flowers on her round-trip hikes for water.

The Thorp lookout rarely saw visitors. The Cle Elum River had to be crossed on horseback without benefit of a bridge. For the packer it was an overnight trip, with a stop at Thorp Lake before the climb to the peak across steep snow slopes on which horses and mules were likely to balk. Joe Ostliff, a retired packer from Roslyn, had to shovel footsteps for them. The mules eyed his handiwork and accepted it, but the horses had to be led.

Thorp is maintained by the Forest Service and staffed by volunteers. Roads have shortened the distance to 2.5 miles.

Goat Peak (near Cle Elum)

Elevation 4,981 feet

Two lookouts have stood on the rounded summit of Goat Peak near Cle Elum, surrounded by acres of beargrass and huckleberries. In season the top is colorful with blossoms of lupine, avalanche lilies, and blue anemones. The lookout had a pleasant view to the south of the forested Big Creek Valley and beyond that to the Cascade Crest, but the sensational view was to the north: up the length of Kachess Lake to Dutch Miller Gap Peaks, and below to farmlands near the town of Easton. At night, car lights could be seen on the highway.

The first lookout, a ground house with cupola, was built about 1920 with materials packed in by horse from Easton. The cupola lookout was replaced in the late 1950s by a modern flat-top on a 12-foot tower. It was used only four or five years before it was declared surplus, subject to destruction. Laurie Contratto, described by his friends as a "lookout buff," bought it and moved the building to his farm up the Teanaway River Road, where it stands today.

Land in this area is part of the railroad land grant checkerboard. At the time the lookout was built it was on Forest Service land—two acres of the peak leased from the Northern Pacific Railroad. When the lease expired, ownership went back to the railroad, now Burlington Northern.

Goat Peak Lookout, now located on a Teanaway farm

Goat Peak (on American Ridge)

Elevation 6,473 feet

Among mountains, Goat Peak is a name like Smith or Jones. This one is a sharp, rocky knoll on 17-mile-long American Ridge between the American River and Bumping River. The view is up the American to Chinook Pass and Mount Rainier, and up the Bumping to the Bumping Lake reservoir and Mount Adams. South is Nelson Ridge and the lookout on Little Bald Mountain.

Directly north, across the American River Valley, are the impressive cliffs of Fifes Peaks, so saw-toothed the mountain had to be named in the plural. Musical instruments had nothing to do with the naming of the multitopped mountain—if they had, Bagpipe Peaks would have been more apt. Tom Fife came from Scotland to homestead in the district in 1887.

The lookout building, a hip-roofed R-6 built in 1934, covered most of the narrow peak, but 20 feet below was a storage shed and plenty of leg-stretching room. It took a bit of leg work to make the round trip to the closest spring, a mile away and almost 1,000 feet lower. The building was removed during the "get-rid-of-them" period of the 1960s.

Goat Peak is right on the American Ridge Trail, a 25.5-mile trail starting near the confluence of the American and Bumping Rivers. With dips and climbs—a lot more ups than downs, it seems—it follows the ridge crest to the boundary of Mount Rainier National Park. A shorter trail starts from Goose Prairie.

Goat Peak Lookout on American Ridge (U.S. Forest Service photo)

Elk below Raven Roost Lookout

Raven Roost

Elevation 6,198 feet

The lookout cabin is not the only thing that has disappeared from Raven Roost in the Naches Ranger District. The mountain itself was lowered by 34 feet to make room for two microwave stations and a parking lot. The panoramic view is still there, dominated by Mount Rainier and the towering needles of Fifes Peaks only 3 miles away; but other details of the scene have changed since the original lookout was erected in 1934.

Sometime before 1900 the huge Chambers fire swept through the area, devastating the land. A 1934 panoramic photograph showed miles of silver snags and meadows, probably the reason elk were seen in bands of 150 to 200. Cattle and sheep also roamed there, providing another chore for the lookout who was supposed to keep track of whether their grazing fees had been paid.

Electronic equipment on the site of Raven Roost Lookout

Native elk had been killed off, so in 1913 forty-two cows and eight bulls were introduced from the Rocky Mountains and granted freedom from pursuit for the next twenty-three years. But by 1936, the elk were indiscreet enough to run around in huge herds of well over a hundred, and annual hunting seasons again were declared.

Today the meadows have been mostly overgrown by young forest, and the elk travel in smaller bands, of twelve to twenty in a group. Dick Simmons, the last person to man the old lookout in 1962 (he had not a single fire to report), said there were still twenty to forty elk grazing near the building.

When the 14-by-14-foot ground building was built, the CCC also constructed a narrow one-way road to it, so it never rated as an isolated station. Nevertheless, the road was so rutted and difficult visitors were seldom seen. Raven Roost didn't become a true drive-in mountaintop until 1964, when it was chopped down to its present altitude. As part of its deal with the Forest Service, AT&T rebuilt the access road and added summit parking, which turned the peak into a minor tourist attraction.

6

OKANOGAN NATIONAL FOREST

As part of the Washington Forest Reserve set aside by President Grover Cleveland in 1897, the Okanogan is one of the oldest national forests in the United States. Until 1955, the entire region was called the Chelan National Forest, clear to the Canadian border from Lake Chelan, and east to the Ferry County line running due north from Grand Coulee.

It covers more than 1.5 million acres, from the Cascade Crest to the Okanogan River Valley, and continues in patches to north of the Colville Indian Reservation. It also laps over the summit of the Cascades down into Whatcom County, to the borders of the Ross Lake National Recreation Area and North Cascades National Park. Technically, the westside land is in the Mount Baker–Snoqualmie National Forest, but as a practical matter it is

The 1934 panorama photo taken from the old cupola lookout at North Twentymile Peak showed a log cabin, which may have been an even earlier lookout. Numbers on the photo indicate landmarks. (U.S. Forest Service photo)

administered by Okanogan. This puts the entire Pasayten Wilderness Area, west and east of the summit, under Okanogan National Forest jurisdiction.

The Pasayten belt across northern Okanogan County is a country of rolling meadows and open woods which always has been more attractive to stockmen than to loggers. From the early part of the century on, great bands of sheep were driven there every summer, and as that industry diminished, beef cattle took their place. The area's few and widely scattered lookouts were nearly as much concerned with grazing permits as they were with fire protection.

Sheep and cattlemen were cagey about revealing exactly how many head of stock they had on a range. They could always hope the Forest Service underestimated and undercharged them. Back in 1936, when sizing up the sheep bands near Remmel Mountain Lookout, Howard Culp came so close to being accurate he surprised the sheepmen.

"How did you arrive at that figure?" they asked in grudging admiration.

Culp pulled an old gag but one that always gets a laugh. "I just counted their feet," he said, "and divided by four."

Remmel was not Culp's only assignment in the Okanogan. Another was Diamond Point, which looked due north along the Ashnola River. Diamond Point came close to being his last assignment. He grounded his telephone as a storm approached one day and watched it pass. After what seemed to be a safe interval, he threw in the switch to report an "all clear." He cranked the telephone . . .

WHAM! One sneaky bolt had been hiding above the cabin. Culp regained consciousness two hours later just as members of a trail crew arrived. The Winthrop Ranger Station had contacted them to go find out why Culp had not been heard from.

Among the now-gone lookouts in the far northern belt was Bunker Hill, a couple of miles south of the Canadian border's Monument 83. A more modern building replaced the original one in the 1960s, just in time to be declared obsolete. It was used only a couple of years before it was closed and burned.

Farther south in the Okanogan, outside today's Pasayten Wilderness, a chain of lookouts stretched across heavily forested country. Watching over the Methow and Chewack River Valleys and their tributaries were lookouts on Goat Peak, Sweetgrass Butte, First Butte, Old Baldy, and Pearrygin. Goat Peak and First Butte were among those used as Aircraft Warning System stations during the Second World War.

Bill Lester wired most of the lookouts for telephone line and lightning protection during the 1920s and 1930s (and after radio became dependable, removed a lot of his own phone installations). His Forest Service career was

Opposite: *Smokejumpers parachuting to a fire on Bunker Hill in 1969*

Knowlton Knob Lookout, 1942 (U.S. Forest Service photo)

interrupted by a year in the hospital and when he returned to work, he took what he called "the best job for recovery." He spent ten seasons as a lookout on First Butte.

He happened to be back there one day in 1970, during the period when lookout buildings were being removed or destroyed. Bob Crandall, who had been the lookout at Monument 83 for a number of years, bought the Sweetgrass building for $10 with the intention of moving it to his own property on Bear Creek. What easier way of doing it than to leave the lookout whole and simply have a logger lift it by crane and load it onto a truck?

As Lester watched from nearby First Butte, the cabin was hoisted high. The cable snapped. Sweetgrass was demolished by a method not used until that moment.

Even when lookouts were to be burned, they occasionally were sold for a token sum to someone who wanted to make a couple of hundred hard-earned dollars by salvaging the copper used primarily in the lightning insulation. It was no consolation to Crandall, therefore, when he ruefully examined the wreckage and discovered that somebody else had already stripped the cabin.

The Okanogan National Forest had an advantage over others in the state whenever it decided to engage in a salvage operation itself. Parachuting to fires was an untested idea when experiments began near Methow in 1939. Actual fire jumps were carried on in the summer of 1940, and the training school in the Methow Valley developed into the North Cascades Smokejumper Base, in operation for fifty years afterwards. When Okanogan lookout buildings were being dismantled, smokejumpers practiced by landing on some of the sites, pulling out equipment, and preparing it for packing out.

Mount Setting Sun

Elevation 7,253 feet

Another mountain on the southern boundary of the Pasayten Wilderness Area is Mount Setting Sun. Like so many others in the Okanogan, it is a high point on a long ridge connecting to other ridges. The lookout on duty watched over the valley of Lost River, which flows into the Methow above Early Winters, and could see down the Methow Valley. On beyond to the southwest, he had a view of the rugged top of 8,676-foot Mount Silver Star.

Toward the end of its existence this station was demoted to what home-insurance companies call an "appurtenant structure," or another building on the same premises. The lookout on Goat Peak, 5 miles south, could not see into the Lost River Valley. So when lightning struck in that area he would trot 5 miles along the connecting ridge to the unoccupied Setting Sun Lookout, see if there were any fires, and then hike back to Goat Peak.

Goat Peak is still in use, but all that remains of the one at Setting Sun are a few rock steps, the usual pile of nails, cables, and broken glass, and an outhouse. A telephone line that ran east along the Setting Sun ridge to the lookout on Sweetgrass Butte was never removed and can still be followed for miles along the rocky ridge.

The trail to the top once started in the Methow Valley near Yellowjacket Creek. It was abandoned when the lookout was removed and is now almost completely lost to logging operations, cattle grazing, and windfalls.

Fog-filled Methow Valley from Mount Setting Sun

Funk Mountain

Elevation 5,122 feet

This mile-high site, on a wooded hill overlooking Conconully, the Salmon River watershed, and rolling range country in the Okanogan Valley, is one of the oldest in the Forest Service. Its history began in 1911 with a tent and a tree to shinny up. A tree topped by a platform still stands close to the present lookout.

The records say a ground house erected in 1932 was the first building, but it is unlikely that tents had provided shelter for twenty-one previous years at such a well-established vantage point. More likely the small hip-roofed building replaced a cabin, but since the house had no cupola, the lookout must have gone on scampering up the old tree when he wanted a better look around.

The original lookout was Harley Heath, who claimed to have the first heliograph in the Northwest. Whether or not he was the first, he became so proficient with that old form of communication—a lamp inside a mirrored box, operated with shutters to blink out Morse code messages—that he was sent around the region to teach other lookouts how to use the instrument.

An old photograph showing the Funk Mountain ground house and the base of the tower that replaced it (U.S. Forest Service photo)

The original tree platform lookout

Three taller peaks in sight of Funk to the west and northwest are Mucka-muck, Old Baldy, and 8,242-foot Tiffany Mountain. Eventually they all had lookouts on them. So did Buck Mountain, with its view north into Con-conully. Most of the land seen from Funk is outside the national forest, and the state operates a couple of lookouts with views up and down the Oka-nogan Valley. Thus, in spite of its long history, in its later years Funk did not enjoy a high priority when there were manpower or budget shortages, and was occupied but sporadically. However, it is still maintained and used during emergencies.

A 1979 photograph showing the old Monument 83 lookout located on the Canadian side of the border and the new tower just inside the United States

Monument 83

Approximate elevation 6,550 feet

Monument 83 is the eighty-third surveying marker on the 49th Parallel, counting inland from Georgia Strait. The lookout was the most re-mote one in Washington, nearly 40 miles by trail through the roadless Pasayten Wilderness Area (though it can now be reached by a 10-mile hike through Canada's Manning Provincial Park). It stands on a high wooded hill that over-looks the Pasayten River Valley in Washington and the tributaries of the Similkameen River in British Columbia. Forests extend in all directions, broken only by an occasional rocky and often snow-covered mountain.

There are two buildings at the site, the older of which is a 12-by-12-foot log cabin with a small sheet-metal cupola. It is probably the only one of its kind, but it is not really in the United States. The cabin was placed by the Forest Service on the highest point of the hill, 125 feet north of the border. Undoubtedly this was done with the approval of the Canadians because the lookout was of equal benefit to them.

A statistics sheet from the old Pasayten Ranger District gives 1930 as the year the first lookout was built, at a cost of $420—surprisingly low consider-ing the remote location. It is generally believed that the log cabin itself dates to the 1920s, and was probably just remodeled in 1930. If only the windows and the sheet metal for the roof and cupola sides had to be packed in, the low cost can be explained.

The hill at Monument 83 is so gradually rounded that a firewatcher could not see over the edge even from the cupola. In 1953 a new lookout was built on a 30-foot wooden tower, this time on the United States side of the border. The old building has been vandalized.

The lookout was closed in 1987, but is still in usable condition. The open knoll furnishes a natural helicopter pad and the Forest Service has permission to land just north of the border, thus avoiding conflict with Pasayten Wilderness rules, which limit helicopter traffic.

Both buildings have frequently been damaged by vandals who pried locks off with a heavy tool. Nothing seems to have been taken, however, and a pack frame from the 1920s was left hanging on a wall. Since the buildings have historic value to both sides, one hopes they will be preserved through international cooperation.

Near 83, another monument marks a grave. Carved on the wooden post is "Pasayten Pete, shot by L. E. Lael 26-2-61." We rather hoped the epitaph would remain an Old West mystery from 1861, but unfortunately the first former Forest Service man we asked about it knew the answer. Bill Lester told us Pasayten Pete was the best-known pack mule in the district, but he had to be shot when he broke a leg in 1961.

Lester also came across the Parson Smith tree when he was 3 miles east of Monument 83, getting logs for the Pasayten Guard Station. Smith was a prospector and trapper of considerable repute as a mountain man. When he stopped in a meadow just south of the border on June 8, 1886, he took time to carve his sentiments on a tree.

> *I have roamed in foreign parts, my boys,*
> *And many lands have seen.*
> *But Columbia is my idol yet,*
> *Of all lands she is queen.*

There was strong support in Washington Territory for joining the Union as the State of Columbia, but three years later, when the step was taken, the name lost out. Bears and rot nearly destroyed the Parson Smith tree so the Okanogan National Forest moved what remained to the Early Winters Visitor's Center. The log was kept in a glass case until the museum's closing in 1996.

The grave of Pasayten Pete

Slate Peak

Elevation 7,440 feet

Slate Peak is on the Pacific Crest Trail but a maintenance road leads to its top. From there you look into the U-shaped valley of the West Fork of the Pasayten River, all timbered; more forest is visible down the deep V-shaped Slate Creek Valley. The biggest show, though, is the whole horizon of peaks—Redoubt, Jack, Baker, Crater, Snowfield, Silver Star, Azurite and, northward, a multitude of rocky ridges in the Pasayten Wilderness Area.

The first lookout was a ground house on the rocky crest of Slate Peak. During World War II it was manned all year long as part of the Aircraft Warning System. Or at least it was used all year—the first winter proved it was virtually impossible to survive in the building during those months. The next winter the man on duty retreated 2 miles to the guard station at Harts Pass to thaw out.

Slate also had lightning problems, and was once set on fire by a direct strike. The bolt melted a keg of nails into one solid mass.

In 1958 or 1959 the Department of Defense sliced off 50 feet of the peak, to make room for a radar station that was to be part of an early warning system. Before the installation was finished, the system was deemed obsolete and was never used.

The Forest Service inherited a good road to the top and a flat area big enough to park a hundred cars. Trouble was, the lookout couldn't sit in a ground house and see over the edges into the valleys anymore. The building had to be raised on a 50-foot tower, bringing it back to the original elevation.

Forest Service packtrain below Slate Peak (U.S. Forest Service photo)

Aerial view of Slate Peak, with Crater Mountain, left, and Jack Mountain, right, on the horizon

It is manned now only during periods of extreme fire danger for radio relay.

Besides scenery, the region has a mining history and an attraction for rock hounds. The ocean covered this part of the globe under a shallow sea at least twice. Mountains rose, fell, and rose again, bringing up with them remnants of sedimentary rock containing marine fossils. The lookout itself is only 2 miles from the old Methow-Chancellor mining road across 6,208-foot Harts Pass.

Mining also followed an off-again-on-again pattern, with short-lived gold rushes in 1859, 1880, and 1887. The longest-lasting one charged into upper Slate Creek, where the town of Barron boomed in the early 1890s. By 1907 most of its miners had given up and departed, leaving the area full of shafts, tunnels, and the remnants of machinery and buildings. At least one mine, though, is still active. Bulldozers razing the surface make it hard to miss.

The old cupola building and the new lookout tower, 1979

North Twentymile Peak

Elevation 7,437 feet

Here is a lookout that should please both the history-minded who want to visit a lookout still standing, and those who just appreciate a sweep-ing view. Hardy backpackers will be able to take its 7 miles of steep trail in stride. Less practiced hikers can gain a feeling of accomplishment from making the climb—after which they can return to just reading about lookouts.

There were three generations of buildings on North Twentymile. Two of the three remain: an example of the 12-by-12 ground house with a 6-by-6 cupola, and a standard hip-roof of the 1930s on a tower. Photographs taken from the cupola in 1930 show the earlier building, a log cabin that was stand-ing at least until then. As at a number of other sites, Forest Service men may have occupied the spot in pre-cabin tents, or "rag houses" as they were called.

According to Forest Service records the cupola type was put up in 1923 at a cost of $796.88. A note added that lightning protection was installed in 1926. Since the same notation appears in records of other buildings, look-outs before 1926 must have had to work on the hope that lightning wouldn't strike once in the same place.

Well over fifty years old, the ground house is a National Historic Build-ing. It is still in good shape. In fact, it was used about 1975 to watch the Black Lake fire, better seen from its cupola than from the newer tower.

The 14-by-14 hip-roof on the 30-foot tower was not built until the mid-1940s, long after most similar designs were popular. There was no rush in erecting it, probably because the ground house was still quite usable. Materials were delivered in 1945 and 1946 and construction went on until 1948, with an end tab of $3,867. To anyone who has built his own home, that kind of schedule should have a familiar ring.

Forest-covered hills stretch from Mount Silver Star in the west to Tiffany Mountain in the east, as well as to hills along the Canadian border to the north and those across the Methow Valley to the south. The only road that can be seen is along the Chewack River more than a mile below. It all appears to be virgin forest but selective logging is under way. Large ponderosa pine and fir are being removed, leaving smaller trees so there are no unsightly clearcuts.

The lookout has not been in use since 1987 but is still in good shape. The 1930 cupola is maintained by volunteers.

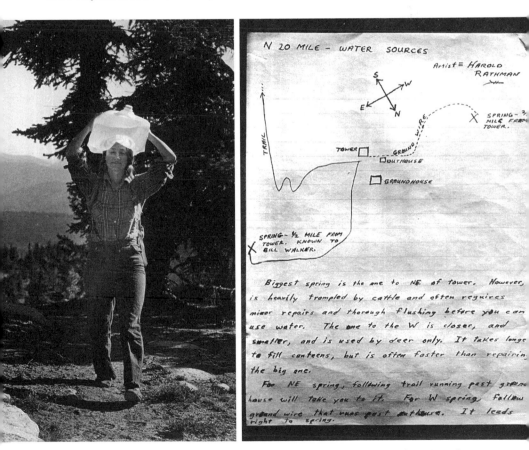

Left, *Lookout Isabelle Spohn carrying water to the lookout in 1979.* Right, *Map of water sources posted on the wall of the lookout building*

Lemanasky (Aeneas Mountain)

Elevation 5,167 feet

The Washington State Department of Natural Resources' lookout on Lemanasky Mountain is an old-timer as state lookouts go, active ever since 1934. The first building was constructed by the CCC as a joint project of the state and the Forest Service. In 1954 there were some land exchanges and the Forest Service sold its interest to the DNR.

As the high point in the long ridge called Aeneas Mountain, Lemanasky has a name of its own. It is covered with sagebrush and dotted with trees. To the east one can see Tonasket and Mount Bonaparte. To the south is the transition between mountains and sagebrush. West is the Sinlahekin Valley and, beyond, the ranges of the North Cascades. Mountains along the border north of the Similkameen also are in sight.

The present lookout was erected in 1980. Like its predecessor, it is a 14-by-14 on a 42-foot tower. The old one began on a 20-foot stand and later was elevated. Like the new one, it had electric lights, refrigerator, and an electric stove as well as a telephone. Lemanasky is a drive-in lookout, but the road is gated to discourage vandals. The road near the top was improved when the building was replaced.

Aeneas Mountain is home to a band of bighorn sheep reintroduced in 1957. During the early 1900s, they had been completely wiped out in the Cascades by disease introduced by domestic sheep, loss of habitat, and over-hunting. They are hunted now on a limited basis, but often enough to make them frightened of humans. However, the lookout occasionally sights them below the tower.

Lemanasky Lookout on Aeneas Mountain. The tower has been replaced since the photograph was taken in 1979.

Panorama photo taken from the top of the original lookout, October 13, 1930 (U.S. Forest Service photo)

Lookout Mountain

Elevation 5,515 feet

This was a case of building the lookout first, then naming the mountain for it. The site, above the town of Twisp, has a broad view of the Methow, Twisp and Libby Creek Valleys. It is much the same view as from the Mount Leecher Lookout, 10 air miles to the southwest, but the two have different perspectives, giving triangulation sightings on fires. Both are still manned, overseeing a complete ring of craggy mountains.

The log cabin at Lookout was built in 1913 by Mac McCowan, on the highest point on Lookout Ridge. A second lookout, on a 30-foot tower, was built by Gus Nelson in 1937. During the Second World War the site was in use during winter as part of the Aircraft Warning System, supplies being packed up with the help of snowshoes. The watchers lived in the log cabin.

From 1938 until the mid-1950s, when the man on Lookout Mountain spoke on the radio he didn't need to identify his station. Everyone recognized the accent of Collin Gillis, and called him "Scotty." Scotty had a reputation for speed and accuracy in reporting fires, but beyond that he also took pride in his immediate domain. It was his yard and he even raked leaves off the trail to the spring a half-mile away. If footprints were left on the path, he erased them, too. He was known as a good cook and Forest Service personnel never carried a lunch when they called at his lookout.

Scotty eventually brought his childhood sweetheart over from Scotland and went to New York to marry her. Any bride should have been happy with a husband so handy in the kitchen and so neat around the place.

Until the mid-1960s the trail to the top was 2.5 miles long, in and out of timber to cross meadows covered with brilliant sunflowers. Since then logging roads have eaten their way up the mountain until the trail itself is only a mile.

Mount Leecher's metal tower and ground house, August 8, 1951 (U.S. Forest Service photo)

Mount Leecher

Elevation 5,012 feet

At least three generations of buildings have been atop this mountain on the dividing line between forest and wheat country, and the present look-out is still in use. In three directions are forested hills, and to the east wheatfields stretch to the horizon. Below are the farms and towns of the Methow Valley, with Carlton and Twisp plainly visible.

According to Mrs. Howard Culp of Okanogan, her uncle and aunt, Dale and Della Price, lived in a tent on Leecher when they were newlyweds. Supplies were packed up by burro. The first recorded building was a ground house put up in 1921 and improved with lightning protection in 1926.

The lookout had a limited view from the cabin, which may account for the ladder leading through the limbs of a topped tree on which is perched a small platform. The tree platform could even outdate the cabin, for it might have been rigged up in tent days. Or it could have been erected after 1921 as the lookout's supplementary tower. In any case, the tree ceded its usefulness in 1932 to a 40-foot steel tower topped by a 6-by-6-foot glass-enclosed platform.

During World War II the ground house was sold to the American Legion and moved to Twisp as a station connected to the Aircraft Warning System.

Its replacement on the mountain was an abandoned lookout brought over from Chiliwist Butte, where it was no longer on Forest Service land due to some title swapping.

In 1953 the steel tower was moved to Intercity Airport between Winthrop and Twisp, to be used as a training tower for the smokejumpers' school, and the Chiliwist cabin on Leecher was hoisted onto a 41-foot tower. With all the building, rebuilding, remodeling, and substitutions, Leecher must hold some kind of record for nonurban renewal.

As a companion station to Lookout Mountain, Leecher got a different slant on the territory up and down the Methow Valley and the lower end of the Chewack. There is a long and very dusty road leading to the lookout. Leecher was closed in 1987 but is still in good condition.

Left, *Abandoned ladder to the original tree platform.* Right, *Present lookout in 1979*

LOOKOUTHOUSES

All the comforts of home were hard to come by in mountaintop buildings reached only by trail. The lookout might be within sight of ice, snow, and tumbling streams but he still lived like a desert rat, lugging water from as far as a mile away and rationing its use. He did the same with firewood, if any was available within reasonable backpacking distance, and if not, he carried up oil for the stove and his lamps.

The comfort that posed the most difficulty of all, though, was the outhouse. It was a rare lookout site that could have obtained a building permit, even under the most lenient of rural interpretations. A solid rock peak not only would fail to pass a septic tank percolation test, it usually offered no soft spots for digging a pit.

Left, *A 1979 photograph of the comfort station on Pugh Mountain.* Right, *A 1979 photograph of the Stampede Pass outhouse*

On Pyramid Mountain the lookout may not have had much privacy, but the view was tremendous.

In many places, adding a toilet was not a simple matter of placing an old-fashioned outhouse at a discreet but handy location. Aside from impenetrable ground, the peak might accommodate only the lookout cabin itself, guywired to keep it there, with the next flat patch a hundred or more feet below. If that, too, was undiggable, the job was to find the nearest ready-made fissure of the right width to put an accessory building over it.

As a result—in contrast to lookout cabins—there were no standard outhouses. They were custom-built to take advantage of whatever individual

Outhouse at Dirtyface Peak, above Lake Wenatchee, as it stood (or leaned) in 1970

site conditions would allow, and that led to architecture and views never dreamed of by Chic Sale, who wrote a book on the subject many years ago. Sale thought he reviewed the topic of outhouses thoroughly, but he had never seen the lookout variety. (Nevertheless, the term "Chic Sale"—referring to an outhouse—then entered the English language.)

Building materials ranged from stone to wood cut from the nearest trees, to lumber or galvanized sheet metal packed in. Some privies provided complete privacy while others were little more than a seat and a roof. A few builders took the challenge tongue-in-cheek and placed seats so they could brag that theirs had the greatest expanse of scenery. Or they made it a two-holer. Never mind that they nor anyone else had to wait for the bathroom; it was a sociable touch.

We once met a fellow who said that when he was a boy, his family had the fanciest outhouse in all Colorado. It was built over an abandoned mine shaft, a thousand feet deep through rose quartz.

We do not know which lookout in Washington State could claim the "pit" depth record, but a number of cliffhangers had hundreds of feet of space below them. Not rose quartz, maybe, but a quarter of a mile of nearly sheer rock, or a glacier down there, was an even more exhilarating amenity.

LOOKOUT LOCATIONS

Numbers on the maps on this and the following pages indicate the location of lookouts that are listed in the Historical Registry on pages 191–210.

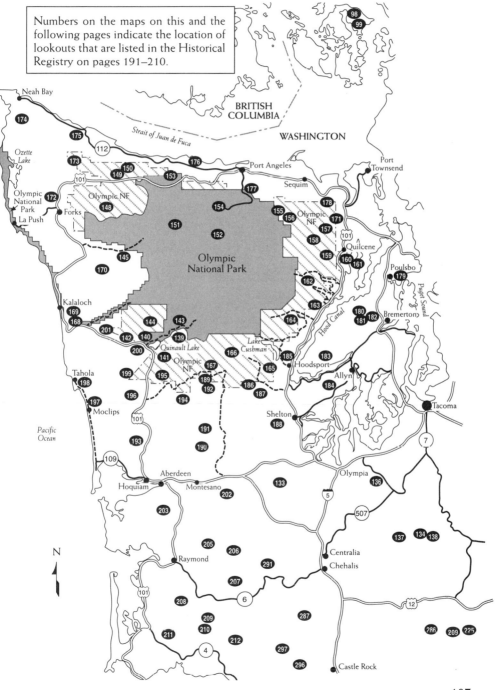

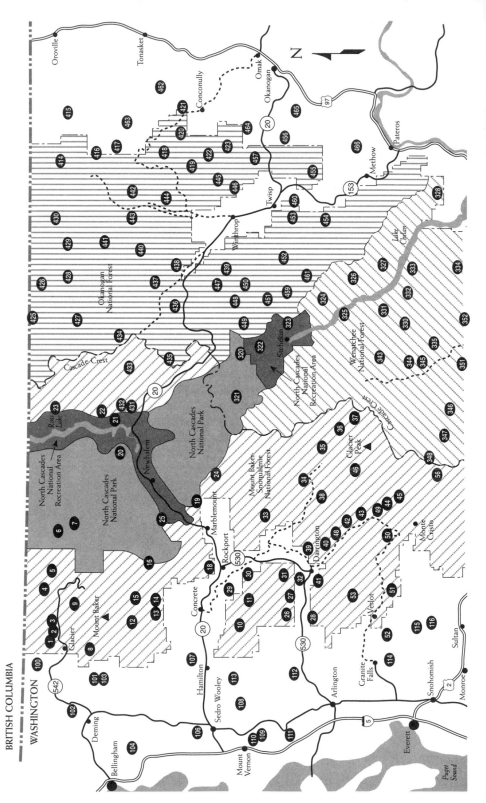

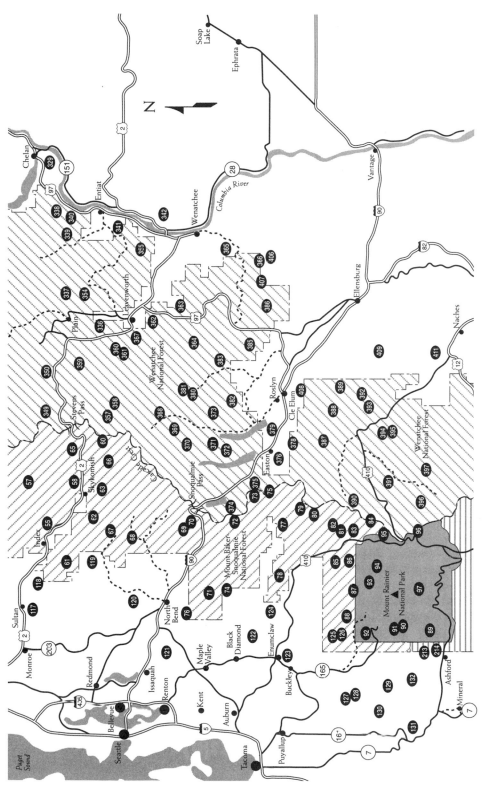

189

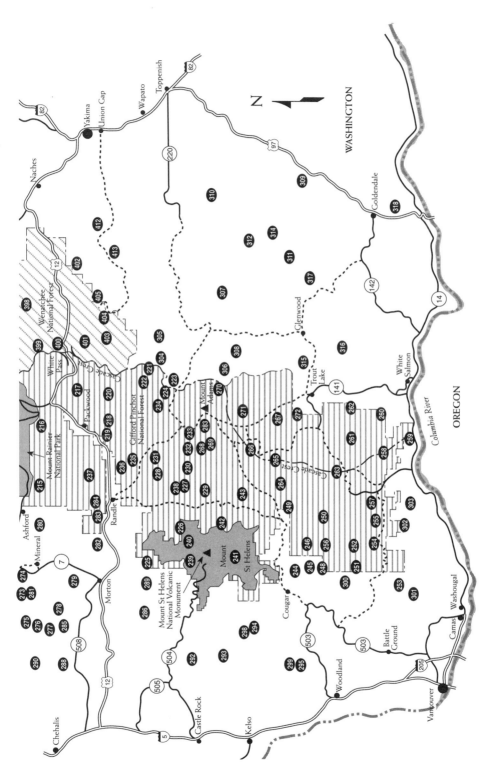

HISTORICAL REGISTRY OF WESTERN WASHINGTON LOOKOUTS

The National Register of Historic Places is a government recognition of the historic significance of a lookout. Getting a lookout so designated is a difficult procedure, but once obtained it provides federal protection. Entry in the Historical Registry of Lookouts is a bit easier to obtain. It gives public recognition to a lookout but extends no official protection. However, it gives the lookout prestige, which helps preserve the structure.

The following is a list of lookouts on the Historical Registry of Lookouts. It was compiled from old maps, interviews, and state and Forest Service records.

While it is intended to be as complete as possible, dates and elevations are often missing, sometimes the locations are unknown, and undoubtedly there were even more lookouts. The publisher and authors would appreciate help in making this list complete. If any reader knows of a lookout not on the list or can help fill in the blanks, please mail your additions to The Mountaineers Books, 1001 SW Klickitat Way, Suite 201, Seattle, WA 98134.

Lookouts in the mountains are best located on a United States Geological Survey (USGS) map, while those in the lowlands are easiest to find on county maps. (Of the several map companies, Metsker maps show the most lookout sites.) To pinpoint a lookout's location, the township and section numbers are included, such as 40-7e-28. The first two numbers refer to township, 40 north-7 east. The third number, 28, is the section. The township and section lines are shown as a grid on most large scale maps.

The access column refers to present conditions and is not necessarily the same as when the lookout was first built. The access does change and anyone wishing to visit one of the sites should get up-to-date directions from the local ranger district.

The following abbreviations are used:

S	Lookout still standing	RT	Restricted Trail
AWS	Used during World War II as a post in the Aircraft Warning System	DT	Difficult Trail
		AT	Abandoned Trail
T	Trail	NT	No Trail

MC	Mountain Climb	SDF	Washington State Department of
R	Road		Forestry, which became the Depart-
RR	Restricted Road		ment of Natural Resources
AR	Abandoned Road	CCC	Civilian Conservation Corps
NRH	National Register of Historic Places	WFFA	Washington Forest Fire Association,
NHLR	National Historic Lookout Register		which became Washington Forest
DNR	Washington State Department of		Protection Association
	Natural Resources		

The abbreviations PRC, CRC, PCL, NL and RC appeared on the DNR computer print-out; however, the people who used the abbreviations have retired and cannot be located. It is assumed the letters refer to a type of tower construction.

No.	Name and Elevation	Map and Location	Built	Removed	Access	Page
	Mount Baker–Snoqualmie National Forest, Baker Ranger District					
1	**West Church Mt.**—5610 ft. No building. Telephone nailed to tree.	USGS Mt. Baker, 40-70-28, 3 mi. NE of Glacier			AT	188
2	**Church Mt.**—6100 ft. Outhouse and storage shed still standing.	USGS Mt. Baker, 40-7E-27, 4 mi. NE of Glacier	1928	1966	T	188
3	**Excelsior Point**—5699 ft. 10×10-ft. building, probably never used.	USGS Mt. Baker, 40-8E-20, 8 mi. NE of Glacier	mid-1930s	1968	T	188
4	**Winchester Mt.**—6521 ft.	USGS Mt. Shuksan, 40-9E-16, 14 mi. NE of Glacier	1935	S	DT	188
5	**Goat Mt.**—4115 ft. 20-ft. tower, manned 1936-1937.	USGS Mt. Shuksan, 40-9E-33, 14 mi. E of Glacier	mid-1930s	c. 1963	AT	188
6	**Copper Mt.**—6260 ft., NRH In North Cascades National Park.	USGS Mt. Challenger, 40-10E-35, 22 mi. E of Glacier	1934	S	T	188
7	**Easy Ridge**—5640 ft. Never used and partly collapsed by snow when removed.	USGS Mt. Challenger, 39-11E-6, 23 mi. E of Glacier	late 1930s	1970	AT	188
8	**Lookout Mt.**—5021 ft. No building. Telephone nailed to tree.	USGS Mt. Baker, 39-7E-35, 4 mi. S of Glacier			AT	188
9	**Barometer Mt.**—5770 ft. No building. Telephone nailed to tree.	USGS Mt. Shuksan, 39-8E-14, 9 mi. SE of Glacier			AT	188
10	**Gee Point**—4974 ft. 10×10-ft. building.	USGS Oso, 34-8E-18, 9 mi. SW of Concrete	1930	1964	R	188
11	**Finney Peak**—5082 ft. 14×14 ft. building.	USGS Finney Peak, 33-8E-2, 11 mi. S of Concrete	1933	1965	R	188
12	**Park Butte**—5450 ft., NRH Maintained by Skagit Alpine Club.	USGS Hamilton, 37-7E-11, 13 mi. NW of Concrete	1936	S	T	188
13	**Dock Butte**—5210 ft.	USGS Hamilton, 36-8E-5, 7 mi. N of Concrete	mid-1930s	1964	T	188
14	**Sulphur Point**—2262 ft. Overlooking Lake Shannon.	USGS Hamilton, 37-8E-27, 9 mi. N of Concrete		c. 1962		188
15	**Dillard Point**—2400 ft. 40-ft. tower currently used to house weather equipment.	USGS Hamilton, 37-8E-22, 10 mi. N of Concrete	1962	1980	R	188
16	**Anderson Butte**—5420 ft. First building 1920-1936, second building 1936-1964.	USGS Lake Shannon, 37-9E-23, 12 mi. NE of Concrete	1920	1964	T	188
17	**Baker Point**	Location unknown	1938	1959		
18	**Sauk Mt.**—5537 ft. Reconstructed in 1957 on small tower.	USGS Lake Shannon, 35-9E-14, 8 mi. E of Concrete	1928	1980	T	188
19	**Lookout Mt.**—5719 ft. Emergency use.	USGS Marblemount, 35-11E-2, 6 mi. NE of Marblemount	late 1920s	S	T	188

No.	Name and Elevation	Map and Location	Built	Removed	Access	Page
20	**Sourdough**—5985 ft., NRH Reconstructed in 1933. Transferred to North Cascades National Park in 1968.	USGS Ross Dam, 38-13E-28, 8 mi. NE of Marblemount	1917	S	T	188
21	**Roland Point**—1983 ft. 35-ft. tower.	USGS Pumpkin Mt., 38-14E-19, 13 mi. NE of Marblemount	mid- 1930s	1959	T	188
22	**Devils Dome**—6982 ft. 14×14-ft. building.	USGS Jack Mt., 39-14E-36, 18 mi. NE of Newhalem	mid- 1930s	1968	T	188
23	**Desolation Peak**—6102 ft., NRH Transferred to North Cascades National Park in 1968.	USGS Hozomeen Mt., 40-14E-32, 20 mi. NE of Newhalem	1932	S	T	188
24	**Hidden Lake Peak**—6890 ft., NRH Maintained by Skagit Alpine Club.	USGS Sonny Boy Lakes, 35-12E-26, 12 mi. E of Marblemount	1931	S	T	188
25	**Bacon Point**—1600 ft. 35-ft. tower.	USGS Marblemount, 36-11E-16, 7 mi. NE of Marblemount	prior to 1934	1956	AT	188

Mount Baker–Snoqualmie National Forest, Darrington Ranger District

No.	Name and Elevation	Map and Location	Built	Removed	Access	Page
26	**Mt. Higgins**—4849 ft. Building badly damaged by snow in 1965. Site first used in 1918.	USGS Oso, 33-8E-32, 10 mi. NW of Darrington	1926		T	188
27	**Round Mt.**—5300 ft. Trail was constructed but lookout may never have been built.	USGS Fortson, 33-8E-27, 8 mi. NW of Darrington			AT	188
28	**French Point**—1100 ft. 14×14-ft. building on 36-ft. tower.	USGS Oso, 32-8E-15, 9 mi. W of Darrington	1934	1959	AR	188
29	**Rinker Point**—3080 ft.	USGS Darrington, 34-9E-35, 11 mi. N of Darrington	mid- 1930s			188
30	**Texas Pond**—approx. 2000 ft. Lookout platform, 1934 building.	USGS Darrington, 33-9E-12, 8 mi. E of Darrington			R	188
31	**North Mt.**—3956 ft. 40-ft. tower, still used.	USGS Darrington, 33-9E-26, 5 mi. N of Darrington	1966	S	R	188
32	**Darrington Ranger Station**—550 ft. Lookout on top of barn. Used 1932 and 1933.	USGS Darrington, 32-9E-14, ½ mi. E of Darrington			R	188
33	**Huckleberry Mt.**—5836 ft. Burned by vandals.	USGS Huckleberry Mt., 33-11E-25, 12 mi. E of Darrington	mid- 1930s	early 1960s	T	188
34	**Green Mt.**—6500 ft., NRH Rebuilt 1931 and 1950.	USGS Downey Mt., 32-12E-3, 17 mi. E of Darrington	1920s	S	T	188
35	**Sulphur Point**—6000 ft. Above Suiattle River. May never have had a building. Used in mid-1930s.	USGS Glacier Peak, 32-13E-28, 20 mi. E of Darrington			T	188
36	**Miners Ridge**—6210 ft., NRH Tent was used until 1930 when a temporary building was constructed. Permanent building built in 1938 and rebuilt in 1953. 40-ft. tower, emergency use.	USGS Glacier Peak, 31-14E-7, 26 mi. E of Darrington	1930	S	T	188
37	**Flower Dome**—6400 ft. 1935 panorama photo. Lookout may not have been built.	USGS Holden, 31-14E-34, 29 mi. SE of Darrington			T	188
38	**Circle Peak**—5983 ft. 14×14-ft. building.	USGS Pugh Mt., 32-11E-36, 13 mi. SE of Darrington	1935	1967	AT	188
39	**Gold Mt.** Maybe a tent. It was the first lookout in Mt. Baker National Forest.	USGS Silverton, 32-10E-32, 4 mi. SE of Darrington	1915	1916	R	188
40	**Dan Creek**—approx. 2000 ft. Used in 1940 and 1941.	USGS White Chuck, 32-10E-27, 5 mi. E of Darrington			S	188
41	**Jumbo Mt.** Probably no building. Used 1916 through the present.	USGS Silverton, 31-9E-11,. 3 mi. S of Darrington			NT	188
42	**White Chuck Bench**—1210 ft. Salvaged, 14×14-ft. building on 40-ft. tower.	USGS White Chuck Mt., 31-11E-29, 15 mi. SE of Darrington	1935	1955	R	188
43	**Pugh Mt.**—7201 ft. Tent in use in 1917, building 1921. The lookout was reached by a rope until trail was built in 1919.	USGS Pugh Mt., 31-11E-27, 13 mi. SE of Darrington	1921	1965	T	188
44	**Red Mt.**—5230 ft. Located on shoulder of Red Mt.	USGS Sloan Peak, 30-12E-28, 20 mi. SE of Darrington	1936	1967	AT	188
45	**Johnson Mt.**—6680 ft.	USGS Bench Mark Mt., 29-13E-18, 27 mi. SE of Darrington	1938	1959	T	188

No.	Name and Elevation	Map and Location	Built	Removed	Access	Page
46	**Glacier Creek Ridge**—5160 ft. 40-ft. tower.	USGS Glacier Peak, 30-13E-31, 22 mi. SE of Darrington	1936	1957	T	188
47	**Suiattle Ridge**	Nothing known about this location				
48	**Black Oak**—2948 ft. A portable tent frame moved to North Mountain in 1960.	USGS White Chuck Mt., 31-11E-7, 15 mi. SE of Darrington	1960		R	188
49	**North Fork Bench**—1281 ft. Salvaged.	USGS Bedal, 30-11E-9, 14 mi. SE of Darrington	1936	1967	R	188
50	**Barlow Pass (Point)**—3400 ft. 14×14-ft cabin.	USGS Bedal, 29-11E-6, 25 mi. E of Granite Falls	mid-1930s	1965	T	188
51	**Blackjack Ridge**—4519 ft. 40-ft. tower near Silverton.	USGS Silverton, 30-9E-26, 17 mi. E of Granite Falls	before 1934	1950	R	188
52	**Mt. Pilchuck**—5324 ft. State park, maintained by Everett branch of The Mountaineers.	USGS Granite Falls, 30-8E-29, 8 mi. E of Granite Falls	1918	S	T	188
53	**Three Fingers**—6854 ft., NRH Maintained by Everett branch of The Mountaineers.	USGS Silverton, 31-9E-18, 7 mi. SW of Darrington	1930	S	MC	188
54	**Verlot Point** A vague reference in Forest Service records. Used for plane spotting during WW II.	Nothing known about this location				

Mount Baker–Snoqualmie National Forest, Skykomish Ranger District

No.	Name and Elevation	Map and Location	Built	Removed	Access	Page
55	**Heybrook**—1701 ft. First a tent with small tower; in 1934 lookout was built on a tower; in 1964 reconstructed to a 14×14-ft. building on a 70-ft. tower. Maintained by Everett branch of The Mountaineers.	USGS Index, 27-10E-21, 14 mi. SE of Sultan	1925 or 1926	S	R	189
56	**Bench Mark Mt.**—5816 ft.	USGS Bench Mark Mt., 28-12E-12, 17 mi. NE of Skykomish	1929-1930	c. 1958	T	188
57	**Evergreen Mt.**—5587 ft., NRH Still used.	USGS Evergreen Mt., 27-12E-16, 10 mi. NE of Skykomish	1935	S	T	189
58	**Beckler Peak**—4950 ft.	USGS Skykomish, 26-12E-20, 3 mi. NE of Skykomish	c. 1924	c. 1950	NT	189
59	**Proffits Point** A low-level lookout near Skykomish.	Location unknown	1935 or 1936	c. 1950	R	
60	**Surprise Mt.**—6330 ft.	USGS Scenic, 25-13E-21, 11 mi. SE of Skykomish	c. 1935	c. 1950	T	189
61	**Mt. Persis** Nothing known but marked on Metsker's Snohomish County map.	USGS Index, 27-9E-34				189
62	**Mt. Cleveland**—5301 ft. First lookout in district, tent with firefinder on wooden tripod. Communicated with heliograph.	USGS Grotto, 25-11E-8, 4 mi. SW of Skykomish	before 1920	1924		189
63	**Maloney Mt.**—3364 ft.	USGS Skykomish, 26-11E-36, 1 mi. SE of Skykomish	1950	1969	R	189
64	**Galena** 14×14-ft. building, near Skykomish, moved to Maloney Mt.	Location unknown	1935-1936	1950		
65	**Windy Mt.**—5386 ft. Tent and tripod used during periods of thunderstorms.	USGS Scenic, 26-14E-20, 9 mi. E of Skykomish	1930s		NT	189
66	**Mt. Sawyer**—5501 ft. Tent and tripod used during periods of thunderstorms.	USGS Scenic, 25-12E-14, 8 mi. SE of Skykomish	1930s		T	189

Mount Baker–Snoqualmie National Forest, North Bend Ranger District

No.	Name and Elevation	Map and Location	Built	Removed	Access	Page
67	**Bare Mt.**—5353 ft. 14×14-ft. building.	USGS Mt. Si, 25-10E-15, 17 mi. NE of North Bend	1935	1973	T	189
68	**Nordrum**—approx. 1500 ft. Near Camp Brown.	USGS Mt. Si, 24-10E-16, 12 mi. E of North Bend			R	189
69	**Lookout Point**—3400 ft. A shelter on the Lake Pratt trail shown on USGS as a lookout.	USGS Bandera, 22-10E-10, 14 mi. SE of North Bend		c. 1960	T	189
70	**Granite Mt.**—5629 ft. First building was a cedar cabin; second building had a cupola; third, rebuilt in 1956 with a 14×14-ft. building. Still used.	USGS Snoqualmie Pass, 22-10E-1, 15 mi. SE of North Bend	before 1920	S	T	189

No.	Name and Elevation	Map and Location	Built	Removed	Access	Page
71	**Little Mt.**—2972 ft. Operated by Seattle in its watershed.	USGS Bandera, 22-9E-21, 10 mi. SE of North Bend			RR	189
72	**Meadow Mt.**—5419 ft. 14×14-ft. building.	USGS Snoqualmie Pass, 21-11E-18, 20 mi. SE of North Bend	1935	1972	R	189
73	**Stampede Pass**—3963 ft. 14×14-ft. building on 35-ft. tower moved to Camp Waskowitz.	USGS Snoqualmie Pass, 22-11E-25, 18 mi. W of Cle Elum	1937	1974	R	189
74	**Humphrey**—2573 ft. 14×14-ft. building on 20-ft. tower	USGS Greenwater, 20-9E-9, 15 mi. E of Enumclaw	1935	1973		189
75	**Snowshoe Butte**—5135 ft.	USGS Lester, 20-11E-14, 20 mi. W of Cle Elum			R	189
76	**Lookout Mt. (Cedar Point)**—2162 ft. On Seattle's watershed; listed as Cedar Point on Forest Service master list.	USGS North Bend, 22-8E-7, 6 mi. SW of North Bend			RR	189
77	**Kelley Butte**—5402 ft. 12×12-ft. building with cupola has been rebuilt.	USGS Lester, 19-10E-2, 24 mi. E of Enumclaw	1950	S	R	189

Mount Baker–Snoqualmie National Forest, White River Ranger District

No.	Name and Elevation	Map and Location	Built	Removed	Access	Page
78	**Christoff (Huckleberry Mt.)**—4764 ft.	USGS Greenwater, 20-9E-36, 17 mi. E of Enumclaw		c. 1965	T	189
79	**Colquhoun Peak**—5715 ft.	USGS Lester, 19-11E-18, 25 mi. E of Enumclaw	early 1960s			189
80	**Pyramid Peak**—5715 ft.	USGS Lester, 19-11E-28, 28 mi. E of Enumclaw	before 1938	1968 or 1969		189
81	**Noble Knob**—6011 ft.	USGS Lester, 18-10E-13, 25 mi. SE of Enumclaw	early 1960s		T	189
82	**Mutton Mt.**—6142 ft. Firefinder manned from Noble Knob 2 mi. away.	USGS Lester, 18-10E-24, 27 mi. SE of Enumclaw			T	189
83	**Three Mile Post** Training lookout, firefinder only. Located at the 3-mi. post on the Corral Pass road.	USGS Lester, 18-10E-35, 26 mi. SE of Enumclaw			R	189
84	**Norse Peak**—6856 ft.	USGS Bumping Lake, 17-11E-18, 30 mi. SE of Enumclaw			T	189
85	**Sun Top**—5271 ft., NRH 14×14-ft. building still in use.	USGS Greenwater, 18-10E-18, 21 mi. SE of Enumclaw	1933	S	R	189
86	**Buck Creek**—5592 ft. Firefinder manned by Suntop 1.5 mi. away. Used in 1930s and 1940s.	USGS Greenwater, 18-10E-30, 22 mi. SE of Enumclaw			T	189
87	**Clear West**—5643 ft.	USGS Greenwater, 18-8E-25, 18 mi. SE of Enumclaw		1968 or 1969	T	189
88	**Bearhead Mt.**—6089 ft.	USGS Enumclaw, 18-8E-29, 15 mi. SE of Enumclaw		early 1960s	T	189

Mount Rainier National Park

No.	Name and Elevation	Map and Location	Built	Removed	Access	Page
89	**Gobblers Knob**—5500 ft., NRH	USGS Mt. Wow, 15-7E-15, SW side of park	1935	S	T	189
90	**Sunset**—5537 ft.	USGS Golden Lakes, 16-7E-11, W side of park	1931	1973	T	189
91	**Colonnade**—approx. 6800 ft.	USGS Mt. Ranier, 16-7E-24, W side of park	1929	1950s	AT	189
92	**Tolmie Peak**—5939 ft., NRH	USGS Golden Lakes, 17-7E-14, NE side of park	1935	S	T	189
93	**Mt. Fremont**—7181 ft., NRH	USGS Sunrise, 17-9E-28, N side of park	1935	S	T	189
94	**Dege Peak**—7006 ft. May have only been a lookout point.	USGS White River Park, 17-10E-30, N side of park			T	189
95	**Crystal Point**—6615 ft. Shown on 1938 Forest Service map.	USGS White River Park, 16-10E-2, N side of park			T	189
96	**Shriner Peak**—5834 ft., NRH Emergency use.	USGS Chinook Pass, 15-10E-3, E side of park	1932	S	T	189
97	**Anvil Rock**—9584 ft. Third-highest lookout in state. Outhouse is still standing.	USGS Mt. Ranier East, 16-9E-31, S side of park	1916	c. 1950	MC	189

No.	Name and Elevation	Map and Location	Built	Removed	Access	Page

Department of Natural Resources, San Juan County

No.	Name and Elevation	Map and Location	Built	Removed	Access	Page
98	**Mt. Constitution**—2409 ft. Stone tower. Now a public viewpoint in Moran State Park.	USGS Mt. Constitution, 37-1W-21, on Orcas island	1936	S	R	187
99	**Little Summit**—2039 ft. 40-ft. tower on Mt. Constitution.	USGS Mt. Constitution, 37-1W-28, on Orcas Island	1966		R	187

Department of Natural Resources, Whatcom County

No.	Name and Elevation	Map and Location	Built	Removed	Access	Page
100	**Black Mt.**—4990 ft. Mentioned in DNR records. It may have been an observation point.	USGS Van Zandt, 40-6E-3, 15 mi. NE of Deming				188
101	**Nooksack**—3350 ft. 50-ft. pole, 14×14-ft. building.	USGS Van Zandt, 38-6E-30, 8 mi. SE of Deming	1957			188
102	**Deming**—2808 ft. Built by SDF, 75-ft. steel tower.	USGS Van Zandt, 39-4E-20, 2 mi. N of Deming	1949			188
103	**Bowman Mt.** On DNR list. May be Nooksack Lookout.	USGS Van Zandt, 38-6E-30, 8 mi. SE of Deming			R	188
104	**Lookout Mt.**—1790 ft. On DNR list.	USGS Bellingham South, 37-3E-11, 4 mi. E of Bellingham			R	188
105	**Entwhistle**—2676 ft. Built by SDF, PRC 40-ft. tower, 14×14-ft. building. Information from DNR computer.	Between Lake Whatcom and Lake Samish	1954			

Department of Natural Resources, Skagit County

No.	Name and Elevation	Map and Location	Built	Removed	Access	Page
106	**Butler Hill**—886 ft. Only foundation left.	USGS Alger, 35-4E-4, 10 mi. N of Mount Vernon			AR	188
107	**Mt. Josephine**—3956 ft. 90-ft. pole, 7×7-ft. building. Rebuilt 1958 with a 14×14-ft. building on a 40-ft. tower.	USGS Hamilton, 36-7E-19, 4 mi. N of Hamilton		1980	R	188
108	**Cultus Mt.**—4089 ft.	USGS Clear Lake, 34-5W-24, 9 mi. E of Mount Vernon			RR	188
109	**Devils Mt.**—1727 ft. Built by CCC. Trees now block all views.	USGS Conway, 33-4E-11, 5 mi. SE of Mount Vernon	1936		R	188
110	**Little Mt.**—934 ft. Now a city park.	USGS Mount Vernon, 34-4E-33, 1 mi. SE of Mount Vernon			R	188
111	**McMurray**—1400 ft. Shown on Metsker map but not in DNR records.	USGS Clear Lake, 33-4E-36, 10 mi. SE of Mount Vernon			R	188
112	**Frailey Mt. (Cavanaugh)**—2795 ft. PRC 40-ft. tower, 14×14-ft. building, built by SDF.	USGS Oso, 33-6E-35, 18 mi. SE of Mount Vernon	1946		RR	188
113	**Little Haystack**—3916 ft.	USGS Clear Lake, 34-6E-8, 13 mi. E of Mount Vernon				188

Department of Natural Resources, Snohomish County

No.	Name and Elevation	Map and Location	Built	Removed	Access	Page
114	**Burn**—735 ft.	USGS Granite Falls, 30-6E-36, 3 mi. S of Granite Falls			R	188
115	**Basin (Sultan)**—2664 ft.	USGS Index, 29-8E-21, 9 mi. NE of Sultan				188
116	**Blue Mt.**—2878 ft.	USGS Monroe, 29-8E-34, 8 mi. NE of Sultan			R	188
117	**High Rock**—1463 ft.	USGS Sultan, 27-7E-23, 3 mi. S of Monroe			AR	189
118	**Haystack**—3590 ft. PRC 40-ft. tower, 14×14-ft. building.	USGS Index, 27-8E-23, 5 mi. SE of Sultan	1957		R	189

Department of Natural Resources, King County

No.	Name and Elevation	Map and Location	Built	Removed	Access	Page
119	**Cascade Peak**—3700 ft. Privately owned.	USGS Mt. Si, 25-9E-5, 15 mi. NE of North Bend		S	GR	189
120	**Snoqualmie (Canyon)**—1691 ft. On Metsker and Forest Service maps.	USGS Lake Joy, 25-8E-9, 12 mi. N of North Bend			AR	189
121	**Tiger Mt.**—3004 ft. 84-ft. tower.	USGS Hobart, 23-7E-8, 9 mi. E of North Bend			R	189
122	**McDonald**—3284 ft. Built by DNR. Not shown on either Metsker or USGS map.	USGS Eagle Gorge, 21-7E-12, 11 mi. NE of Enumclaw	1963		RR	189

No.	Name and Elevation	Map and Location	Built	Removed	Access	Page
123	**Pinnacle Peak (Mt. Pete, Mt. Peak)**—1801 ft. Now a city park.	USGS Enumclaw, 20-7E-31, 2 mi. SE of Enumclaw			T	189
124	**Grass Mt.**—4390 ft. 50-ft. tower and 14×14-ft. building built by CCC. Rebuilt in 1958.	USGS Enumclaw, 20-8E-21, 8 mi. E of Enumclaw	1936	1980	R	189

Department of Natural Resources, Pierce County

No.	Name and Elevation	Map and Location	Built	Removed	Access	Page
125	**South Prairie (O'Farrell)**—3235 ft. Built and manned by the Forest Service. 50-ft. tower.	USGS Enumclaw, 18-7E-8, 10 mi. S of Enumclaw		early 1960s	R	189
126	**Carbon**—4144 ft. Replaced South Prairie (O'Farrell).	USGS Enumclaw, 18-7E-17, 11 mi. S of Enumclaw	early 1960s		R	189
127	**Electron**—2435 ft.	USGS Wilkeson, 18-6E-30, 13 mi. NE of Eatonville			RR	189
128	**McGuire Creek**—approx. 2500 ft.	USGS Kapowsin, 17-6E-3, 14 mi. NE of Eatonville			RR	189
129	**St. Paul**—2973 ft. Built by SDF, PRC 40-ft. tower.	USGS Kapowsin, 17-6E-29, 10 mi. NE of Eatonville	1954		RR	189
130	**Ohop**—2335 ft. Built by CCC, CRC 53-ft. tower, 14×18-ft. building.	USGS Kapowsin, 16-5E-6, 3 mi. NE of Eatonville	1937		R	189
131	**Pack Forest**—2034 ft.	USGS Eatonville, 16-4E-34, 3 mi. S of Eatonville.	1929	S	R	189
132	**Puyallup Ridge**—4930 ft. DNR tower. Seriously damaged 1990.	USGS Kapowsin, 16-6E-32, 10 mi. W of Eatonville	1954	1980	RR	189

Department of Natural Resources, Thurston County

No.	Name and Elevation	Map and Location	Built	Removed	Access	Page
133	**Capitol Peak**—2658 ft.	USGS Rochester, 17-4W-11, 12 mi. SW of Olympia			R	187
134	**Porcupine Ridge**—2252 ft. Built by CCC, CRC 100-ft. tower, 7×7-ft. building.	USGS Yelm, 15-2E-15, 23 mi. SE of Olympia	1956		RR	187
135	**Deschutes**—600 ft. Built by Army, CRC 60-ft. tower, 7×7-ft. building on Fort Lewis Reserve. Found on DNR computer.	Location unknown.	1955		RR	
136	**Rainier Tower**—520 ft. Built by Army on Fort Lewis Reserve. This may be another name for Deschutes.	USGS Yelm, 17-1E-4, 10 mi. E of Olympia			RR	187
137	**Miller Hill**—1828 ft. Shown on USGS map.	USGS Yelm, 15-1E-10, 19 mi. SE of Olympia			R	187
138	**Clam Mt.**—2725 ft. Shown on USGS map as Weyerhaeuser Lookout.	USGS Yelm, 15-2E-22, 25 mi. SE of Olympia			R	187

Olympic National Forest and Olympic National Park, Quinault Ranger Districts

No.	Name and Elevation	Map and Location	Built	Removed	Access	Page
139	**Colonel Bob**—4492 ft. Forest Service lookout; partly destroyed by snow, then burned.	USGS Grisdale, 23-8W-7, 8 mi. E of Amanda Park	1932	1967	T	187
140	**Higley Peak**—3025 ft. Forest Service lookout, AWS, rebuilt 1964	USGS Kloochman Rock, 23-10W-1, 4 mi. N of Amanda Park	1932	1973	R	187
141	**Chester Ridge** Lookout shown on Forest Service map.	USGS Quinault Lake, 22-9W-14, 7 mi. SE of Amanda Park			R	187
142	**Raft** Lookout shown on Forest Service map and not on USGS.	USGS Salmon River, 23-11W-12, 10 mi. NW of Amanda Park			R	187
143	**Finley Creek**—3419 ft. First lookout in Olympic National Forest. Although the lookout was named after the creek, it is assumed to have been on or near the top of Finley Peak. Now in the national park.	USGS Kloochman Rock, 24-9W-12, 12 mi. NE of Amanda Park	1915	1947	AT	187
144	**Kloochman Rock**—3356 ft. Forest Service lookout.	USGS Kloochman Rock, 25-10W-14, 16 mi. N of Amanda Park			T	187

Olympic National Forest and Olympic National Park, West and Forks Ranger Districts

No.	Name and Elevation	Map and Location	Built	Removed	Access	Page
145	**Huelsdonk**—approx. 500 ft. Nothing known. Shown on 1930 Kroll map. Forest Service land.	USGS Spruce Mt., 27-11W-36, 18 mi. SE of Forks				187
146	**Burma Point** Lookout point used in 1940s and 1950s by Forks Ranger.	USGS Forks, 28-13W-?, near Forks Exact location unknown			R	

No.	Name and Elevation	Map and Location	Built	Removed	Access	Page
147	**Ice Cream Cone** Lookout point used in 1940s and 1950s by Forks Ranger.	USGS Forks, 28-13W-?, near Forks Location unknown		R		
148	**Hyas**—approx. 3000 ft. Forest Service lookout.	USGS Spruce Mt., 29-11W-33, 12 mi. E of Forks	1950	R		187
149	**Kloshe Nanich**—approx. 3000 ft. Forest Service lookout removed when North Point was built.	USGS Pysht, 30-11W-23, 18 mi. NE of Forks	1920s	R		187
150	**North Point**—3340 ft. Forest Service lookout, badly vandalized.	USGS Pysht, 30-11W-23, 19 mi. NE of Forks	late 1930s	R		187
151	**Bogachiel Peak**—5474 ft. National Park lookout.	USGS Bogachiel Peak, 28-9W-36, 22 mi. SW of Port Angeles	late 1920s	T		187

Olympic National Park, East District

No.	Name and Elevation	Map and Location	Built	Removed	Access	Page
152	**Dodger Point**—5753 ft. Trail is seldom maintained. Still standing.	USGS Mt. Olympus, 28-7W-36, 17 mi. S of Port Angeles	S	T		187
153	**Pyramid Mt.**—3000 ft. Probably used only for airplane spotting during WW II. Still standing.	USGS Lake Crescent, 30-9W-22, 17 mi. W of Port Angeles	1980	T		187
154	**Hurricane Hill**—5757 ft.	USGS Mt. Olympus, 29-7W-23, 9 mi. SW of Port Angeles		T		187
155	**Deer Park (Blue Mt.)**—6007 ft.	USGS Mt. Angeles, 28-5W-2, 12 mi. SE of Port Angeles	1931	R		187

Olympic National Forest, Quilcene Ranger District

No.	Name and Elevation	Map and Location	Built	Removed	Access	Page
156	**Ned Hill**—3450 ft. Tower but no building. Still standing.	USGS Tyler Peak, 29-4W-34, 14 mi. SE of Port Angeles	1934	S	AT	187
157	**Mt. Zion**—4273 ft. 16×16-ft. building on the ground.	USGS Tyler Peak, 28-3W-13, 10 mi. NW of Quilcene	1929	1975	T	187
158	**Mt. Townsend**—6212 ft. 16×16-ft. building on the ground.	USGS Tyler Peak, 28-3W-33, 9 mi. W of Quilcene		T		187
159	**Big Quilcene Ridge**—approx. 4000 ft. Foundation left, no information.	USGS Mt. Walker, 27-2W-29, 5 mi. SW of Quilcene		R		187
160	**Mt. Walker (north)**—2804 ft. Ground building.	USGS Mt. Walker, 27-2W-36, 3 mi. SW of Quilcene		R		187
161	**Mt. Walker (south)**—2759 ft. No building. Viewpoint used by Mt. Walker North lookout.	USGS Mt. Walker, 27-2W-36, 3 mi. SW of Quilcene	1931	1966	R	187

Olympic National Forest, Hoodsport Ranger District

No.	Name and Elevation	Map and Location	Built	Removed	Access	Page
162	**Mt. Jupiter**—5701 ft. Ground building used as AWS in WW II.	USGS The Brothers, 26-3W-33, 12 mi. SW of Quilcene	c. 1969	T		187
163	**Webb**—2775 ft. Looked like a square water tower.	USGS The Brothers, 25-3W-35, 16 mi. SW of Quilcene	1930	1967	R	187
164	**Jefferson Ridge Point**—3832 ft. 54-ft. tower.	USGS The Brothers, 24-4W-11, 13 mi. N of Hoodsport	1961	1967	T	187

Olympic National Forest, Shelton Ranger District

No.	Name and Elevation	Map and Location	Built	Removed	Access	Page
165	**Dennie Ahl**—2004 ft. Shared with state.	USGS Mt. Tebo, 22-5W-13, 6 mi. W of Hoodsport	1930s	1968	R	187
166	**Dusk Point**—3100 ft. Shown on Forest Service map.	USGS Mt. Tebo, 22-6W-8, 16 mi. W of Hoodsport		R		187
167	**Anderson Butte**—3358 ft.	USGS Grisdale, 22-7W-22, 20 mi. W of Hoodsport	1930s	1968	AT	187

Department of Natural Resources, Jefferson County

No.	Name and Elevation	Map and Location	Built	Removed	Access	Page
168	**Elk Creek**—490 ft. Quinault Indian Reservation.	USGS Destruction Island, 24-13W-25, 15 mi. NW of Amanda Park				187
169	**Clearwater**—705 ft.	USGS Destruction Island, 24-13W-14, 16 mi. NW of Amanda Park				187
170	**Octopus**—2486 ft. PRC 40-ft. tower, 14×14-ft. building, Standing in 1974.	USGS Spruce Mt., 26-11W-19, 17 mi. SE of Forks	1957	R		187
171	**Skidder Hill**—2126 ft. PRC 40-ft. tower, 14×14-ft. building.	USGS Uncas, 28-2W-3, 9 mi. N of Quilcene	1957			187

No.	Name and Elevation	Map and Location	Built	Removed Access	Page

Department of Natural Resources, Clallam County

No.	Name and Elevation	Map and Location	Built	Removed Access	Page
172	**Gunderson Hill**—1235 ft. Built by SDF, 75-ft. steel tower, 8×8-ft. building.	USGS Lake Pleasant, 29-13W-20, 4 mi. N of Forks	1948	R	187
173	**Ellis Mt.**—2338 ft. Built by SDF, 14×14-ft. ground house.	USGS Lake Pleasant, 30-13W-1, 13 mi. N of Forks	1952	R	187
174	**Sooes**—1978 ft. Built by Forest Service, 90-ft. steel tower. 8×8-ft. building on Makah Indian Reservation.	USGS Cape Flattery, 32-15W-11, 14 mi. E of Sekiu	1957	RR	187
175	**Sekiu**—approx. 1500 ft. On Forest Service computer.	USGS Lake Pleasant, 31-14W-2, 5 mi. S of Sekiu			187
176	**Striped Peak**—1015 ft. Built by CCC, CRC 53-ft. tower.	USGS Joyce, 31-8W-27, 10 mi. W of Port Angeles	1940		187
177	**Mt. Pleasant**—2638 ft. Built by SDF, PRC 40-ft. tower, 14×14-ft. building. New tower 1964.	USGS Port Angeles, 30-6W-31, 4 mi. S of Port Angeles	1953		187
178	**Mt. Blyn**—1966 ft. State lookout on Forest Service land.	USGS Gardiner, 29-2W-18, 22 mi. SE of Port Angeles			187

Department of Natural Resources, Kitsap County

No.	Name and Elevation	Map and Location	Built	Removed Access	Page
179	**Lookout**—440 ft. On Metsker map near Indianola.	USGS Port Gamble, 26-2E-7, 13 mi. N of Bremerton			187
180	**Green Mt.**—1780 ft. Built by CCC, CRC 84-ft. tower, 14×14-ft. building. Rebuilt 1963, 50-ft. tower, 14×14-ft. building.	USGS Wildcat Lake, 24-1W-21, 7 mi. W of Bremerton	1942	R	187
181	**Gold Mt.**—1560 ft. Built by SDF, 75-ft. steel tower, 8×8-ft. building. New tower 1965.	USGS Wildcat Lake, 24-1W-27, 6 mi. W of Bremerton	1948	RR	187
182	**(no name)**—1380 ft. Shows only on USGS map.	USGS Bremerton West, 24-1W-25, 3 mi. W of Bremerton			187

Department of Natural Resources, Mason County

No.	Name and Elevation	Map and Location	Built	Removed Access	Page
183	**Tahuya**—443 ft.	USGS Lake Wooten, 23-2W-22, 10 mi. NE of Hoodsport		R	187
184	**Mason Lake**—362 ft.	USGS Mason Lake, 22-2W-34, 12 mi. SE of Hoodsport		R	187
185	**Dow Mt.**—2500 ft. Built by WFFA, 10×10-ft. smoke house. Rebuilt 1963 on 40-ft. tower.	USGS Potlatch, 23-4W-34, 4 mi. NW of Hoodsport	1942	AR	187
186	**South Mt.**—3100 ft. Built by SDF, 14×14-ft. ground house.	USGS Mt. Tebo, 21-5W-18, 12 mi. SW of Hoodsport	1956	R	187
187	**Simpson**—1200 ft.	USGS Mt. Tebo, 21-5W-22, 11 mi. NW of Shelton		R	187
188	**Dayton Peak**—920 ft. Built by SDF, 10×10-ft. ground house with cupola. Moved 1963 0.25 mile and rebuilt on tower.	USGS Shelton, 20-4W-30, 10 mi. W of Shelton		R	187

Department of Natural Resources, Grays Harbor

No.	Name and Elevation	Map and Location	Built	Removed Access	Page
189	**Weatherwax**—2640 ft. Built by SDF, PRC 40-ft. tower, 14×14-ft. building.	USGS Grisdale, 21-7W-6, 23 mi. NE of Aberdeen	1956	R	187
190	**Prices Peak**—730 ft. Built by SDF, PRC 40-ft. tower, 14×14-ft. building.	USGS Wynoochee Valley, 19-7W-32, 13 mi. NE of Aberdeen	1954	R	187
191	**Mobray**—656 ft.	USGS Wynoochee Valley, 20-8W-36, 17 mi. NE of Aberdeen		R	187
192	**Drake (Reed Hill)**—1443 ft.	USGS Grisdale, 21-7W-20, 24 mi. NE of Aberdeen		R	187
193	**Deer Creek**—605 ft. Built by SDF, PRC 40-ft. tower, 14×14-ft. building.	USGS Humptulips, 19-10W-20, 10 mi. NW of Aberdeen	1954	R	187
194	**Brittain**—1228 ft.	USGS Quinault Lake, 21-9W-36, 19 mi. N of Aberdeen		R	187
195	**Burnt Hill**—1000 ft. Built by Forest Service, smoke house, 14×14-ft. building.	USGS Quinault Lake, 21-9W-7, 25 mi. N of Aberdeen		R	187

No.	Name and Elevation	Map and Location	Built	Removed	Access	Page
196	**Macafee**—655 ft.	USGS Macafee Hill, 21-11W-24, 23 mi. NW of Aberdeen			R	187
197	**Moclips** Does not show on either USGS Macafee Hill or Moclips.	USGS Moclips, 20-12W-5, 26 mi. NW of Aberdeen			R	187
198	**Point Grenville**—516 ft. Quinault Indian Reservation.	USGS Macafee Hill, 21-12W-17, 28 mi. NW of Aberdeen			RR	187
199	**Pete Miller's Treehouse**—264 ft. **(officially Cook Creek)** Built on top of a tree on the Quinault Indian Reservation.	USGS Macafee Hill, 22-11W-26, 9 mi. SW of Amanda Park	1927	1940	RR	187
200	**Lone Mt.**—1173 ft. Quinault Indian Reservation.	USGS Quinault Lake, 23-10W-20, 4 mi. W of Amanda Park			RR	187
201	**Salmon River**—2693 ft. Trailer 1966-1967.	USGS Salmon river, 23-11W-4, 12 mi. NW of Amanda Park			R	187
202	**Mt. Minot**—1768 ft. Built by WFFA, PRC 40-ft. tower, 14×14-ft. building.	USGS Malone, 16-6W-10, 19 mi. SE of Aberdeen	1935		R	187
203	**Artic**—700 ft. Built by SDF, PRC 40-ft. tower, 14×14-ft. building.	USGS Aberdeen, 16-9W-14, 5 mi. S of Aberdeen	1953		R	187
204	**Twin Peak** On DNR computer.	Location unknown				

Department of Natural Resources, Pacific County

No.	Name and Elevation	Map and Location	Built	Removed	Access	Page
205	**Pack Sack**—1423 ft. 50-ft. pole, 14×14-ft. building, built by SDF.	USGS Raymond, 15-7W-33, 9 mi. NE of Raymond	1950		R	187
206	**Burt**—1384 ft. Built by SDF.	USGS Pe Ell, 14-6W-22, 18 mi. E of Raymond			R	187
207	**Squally Jim**—2810 ft. PRC 40-ft. tower built by SDF.	USGS Pe Ell, 13-6W-28, 16 mi. SE of Raymond	1955		R	187
208	**Trap Creek**—2152 ft. Built by WFFA, 14×14-ft. ground house. Relocated on 20-ft. tower in 1963.	USGS Raymond, 12-8W-7, 10 mi. S of Raymond	1942		R	187
209	**K.O. Point**—2632 ft. PRC 20-ft. tower, 14×14-ft. building.	USGS Grays River, 11-7W-4, 18 mi. SE of Raymond	1962		R	187
210	**Hull Creek**—2042 ft. On Metsker map.	USGS Grays River, 11-7W-8, 19 mi. SE of Raymond			R	187
211	**Cowan Peak**—1940 ft. Built by WFFA, NL 40-ft. tower.	USGS Oman Ranch, 11-9W-27, 22 mi. S of Raymond	1953		R	187
212	**Blaney**—2546 ft. Built by SDF, PRC 84-ft. tower, 14×14-ft. building. Replaced building in 1960.	USGS Skamokawa, 11-6W-23, 26 mi. SE of Raymond	1960		R	187

Gifford Pinchot National Forest, Packwood Ranger District

No.	Name and Elevation	Map and Location	Built	Removed	Access	Page
213	**Glacier View**—5450 ft.	USGS Mt. Wow, 15-7E-5, 6 mi. NE of Ashford	1930s	1960s	T	189
214	**Mt. Beljica**—5475 ft. No lookout building. Tool shed, still standing. Telephone was nailed on a tree.	USGS Mt. Wow, 15-7E-17, 5 mi. NE of Ashford			AT	189
215	**High Rock**—5685 ft.	USGS Randle, 14-7E-22, 8 mi. SE of Ashford	1930s	S	T	190
216	**Tatoosh**—6310 ft.	USGS Packwood, 14-9E-15, 7 mi. N of Packwood	1920s	1960s	T	190
217	**Lost Lake**—6359 ft.	USGS Packwood, 13-10E-24, 9 mi. E of Packwood	1930s	1960s	T	190
218	**South Point**—5980 ft. On tower.	USGS Packwood, 12-9E-23, 6 mi. S of Packwood	1932	1972	T	190
219	**Dry Creek**—3815 ft. On tower.	USGS Packwood, 12-9E-8, 6 mi. S of Packwood	1935	1966	T	190
220	**Hawkeye Point**—7431 ft. Used only four years.	USGS White Pass, 12-10E-24, 11 mi. SE of Packwood	1927	1966	T	190
221	**Nannie Peak**—6106 ft.	USGS Walupt, 11-11E-10, 16 mi. SE of Packwood	c. 1934	1960s	T	190
222	**Goat Ridge**—6240 ft.	USGS Hamilton Butte, 12-10E-35, 12 mi. SE of Packwood	1933	1960s	T	190
223	**Midway**—5236 ft. Operated by Midway Guard Station.	USGS Green Mountain, 10-10E-12, 20 mi. SE of Packwood	1934		R	190

No.	Name and Elevation	Map and Location	Built	Removed	Access	Page
224	Lookout Mountain (Angry Mountain)—5245 ft. On Forest Service records.	USGS Packwood Lake, 12-10E-17,			T	190

Gifford Pinchot National Forest, Randle Ranger District

No.	Name and Elevation	Map and Location	Built	Removed	Access	Page
225	Vanson Peak—4948 ft.	USGS Spirit Lake, 11-5E-26, 13 mi. SW of Randle		1960s	T	190
226	Strawberry Mt.—5464 ft.	USGS Spirit Lake, 10-6E-22, 15 mi. S of Randle		1960s	T	190
227	French Butte—4607 ft.	USGS French Butte, 10-7E-22, 14 mi. S of Randle	1929	1960s	R	190
228	Burley Mt.—5310 ft. Badly vandalized in 1978.	USGS Tower Rock, 11-7E-25, 10 mi. SE of Randle	1932	S	R	190
229	Badger Mt.—5664 ft.	USGS French Butte, 9-7E-11, 19 mi. S of Randle	1926	1960s	T	190
230	McCoy Peak—5849 ft.	USGS McCoy Peak, 10-8E-20, 16 mi. SE of Randle	1934	1960s	T	190
231	Tongue Mt.—4838 ft. Lookout was gone by 1948.	USGS Tower Rock, 11-8E-26, 13 mi. SE of Randle	1934		AT	190
232	Sunrise Peak—5880 ft. Shown on 1931 Forest Service map.	USGS McCoy Peak, 10-8E-24, 18 mi. SE of Randle	1934	1960s	T	190
233	Council Bluff (Butte)—5180 ft.	USGS East Canyon Ridge, 9-9E-11, 24 mi. S of Packwood	1932	1960s	T	190
234	Hamilton Butte—5772 ft.	USGS Hamilton Butte, 11-10E-31, 16 mi. S of Packwood	1929	1960s	R	190
235	Cispus (Twin Sisters)—5647 ft. A cabin, built in 1915, 0.5 mile from lookout site still standing.	USGS Tower Rock, 11-9E-5, 13 mi. SE of Randle	1926	1960s	T	190
236	Pompey Peaks—5177 ft.	USGS Randle, 12-8E-26, 10 mi. E of Randle	1934	1960s	T	190
237	Trails End (Purcell Mt.)—5442 ft.	USGS Randle, 13-8E-29, 17 mi. NE of Randle	1933	1960s	T	190
238	Ferrous Point—4800 ft. On Metsker map. Probably serviced by French Butte Lookout.	USGS French Butte, 10-7E-16, 13 mi. S of Randle			R	190

Gifford Pinchot National Forest, St. Helens Ranger District

No.	Name and Elevation	Map and Location	Built	Removed	Access	Page
239	Coldwater—5727 ft.	USGS Spirit Lake, 10-5E-33, 21 mi. SW of Randle			T	190
240	Mt. Margaret—5858 ft. May have only been an observation point.	USGS Spirit Lake, 10-5E-36, 19 mi. SW of Randle			T	190
241	Mt. St. Helens—9677 ft. Building had collapsed in the 1940s, but some boards remained until the 1980 eruptions.	USGS Mt. St. Helens, 8-5E-9, 11 mi. NE of Cougar	1922	1980	MC	190
242	Smith Creek Butte—3820 ft.	USGS Mt. St. Helens, 8-6E-3, 17 mi. NE of Cougar			R	190
243	Spencer Butte—4247 ft. Shown on 1931 Forest Service map.	USGS Spencer Butte, 8-7E-20, 20 mi. NE of Cougar	prior to 1931		T	190
244	Mt. Mitchell—3926 ft.	USGS Mt. St. Helens, 6-5E-4, 6 mi. E of Cougar				190
245	Huffman Peak—4106 ft.	USGS Lookout Mt., 6-5E-27, 9 mi. SE of Cougar			T	190
246	Siouxon Peak—4169 ft.	USGS Lookout Mt., 6-5E-24, 9 mi. SE of Cougar			R	190
247	Switchback A firefinder located on a switchback of a forest road.	Location unknown.			R	

Gifford Pinchot National Forest, Wind River Ranger District

No.	Name and Elevation	Map and Location	Built	Removed	Access	Page
248	West Point—3850 ft. On Forest Service list.	USGS Lookout Mt., 5-5E-10, 11 mi. SE of Cougar			R	190
249	Point 3670—3670 ft. Building collapsed.	USGS Burnt Peak, 7-7E-31, 16 mi. E of Cougar		1960s	R	190
250	Observation Peak (Trapper)—4207 ft. Second building on nearby Sister Rocks.	USGS Lookout Mt., 5-6E-14, 18 mi. N of Stevenson			T	190

No.	Name and Elevation	Map and Location	Built	Removed	Access	Page
251	**Gumboot Mt.**—3856 ft.	USGS Lookout Mt., 4-5E-5, 20 mi. NW of Stevenson				190
252	**Green Lookout Mt.**—4442 ft.	USGS Lookout Mt., 4-5E-1, 16 mi. NW of Stevenson			R	190
253	**Silver Star Mt.**—4390 ft.	USGS Bridal Veil, 3-5E-18, 18 mi. W of Stevenson			R	190
254	**Lookout Mt.**—4222 ft.	USGS Lookout Mt., 4-6E-29, 13 mi. NW of Stevenson			R	190
255	**Mowich Butte**—3513 ft.	USGS Lookout Mt., 4-6E-26, 10 mi. NW of Stevenson			R	190
256	**Horseshoe Ridge**—3495 ft.	USGS Lookout Mt., 5-6E-7, 12 mi. SE of Cougar			T	190
257	**Bunker Hill**—2350 ft.	USGS Wind River, 4-7E-22, 9 mi. N of Stevenson				190
258	**Grassy Knoll**—3649 ft.	USGS Wind River, 4-8E-23, 7 mi. W of Willard			T	190
259	**Puppy (Dog Mt.)**—approx. 2400 ft. Overlooked Columbia River.	USGS Hood River, 3-9E-29, 6 mi. SW of Willard			T	190
260	**Little Baldy Peak**—2924 ft.	USGS Willard, 4-9E-25, 1 mi. N of Willard			R	190
261	**Little Huckleberry Mt.**—4781 ft.	USGS Willard, 5-9E-29, 8 mi. N of Willard			T	190
262	**Monte Cristo**—4171 ft.	USGS Willard, 5-10E-20, 10 mi. NE of Willard			T	190

Gifford Pinchot National Forest, Mt. Adams Ranger District

No.	Name and Elevation	Map and Location	Built	Removed	Access	Page
263	**Red Mt.**—4968 ft. Staffed by volunteers 1994-1995. Forest Service photograph caption reads 1910. Road is steep and sometimes gated.	USGS Wind River, 5-8E-8, 14 mi. NW of Willard	S		R	190
264	**Burnt Peak**—4135 ft.	USGS Burnt Peak, 7-7E-23, 18 mi. W of Trout Lake	1931		AT	190
265	**West Twin Butte**—4716 ft. 12×12-ft. ground building replaced in 1930s by cabin on a 12-ft. tower.	USGS Lone Butte, 7-8E-2, 14 mi. NW of Trout Lake	1923	1963	R	190
266	**Steamboat Mt.**—5425 ft. Rebuilt 1956.	USGS Steamboat Mt., 8-9E-31, 13 mi. NW of Trout Lake	1927	1971	T	190
267	**Sleeping Beauty**—4907 ft.	USGS Sleeping Beauty, 7-9E-14, 9 mi. NW of Trout Lake	1931		T	190
268	**Summit Prairie**—5238 ft. Last used as a lookout in 1936. Used as a tool shed until 1967.	USGS East Canyon Ridge, 9-9E-18, 21 mi. NW of Trout Lake	1929 or 1930	1967	T	190
269	**Dark Mt. Trail**—approx. 4900 ft. Cedar cabin used before Summit Prairie Lookout.	USGS East Canyon Ridge, 9-9E-18, 21 mi. NW of Trout Lake			T	190
270	**Mt. Adams**—12,276 ft. Abandoned 1924 but maintained until 1959 by sulfur miners. Since then buried in summit ice cap.	USGS Mt. Adams East, 10-8E-12, 14 mi. N of Trout Lake	1918-1921	S	MC	190
271	**Madcat Meadow**—approx. 5800 ft. Portable firefinder and shed. Collapsed remains can be seen.	USGS Mt. Adams West, 10-8E-22, 12 mi. N of Trout Lake	late 1920s		T	190
272	**Flattop Mt.**—4394 ft. Forest Service lookout on state land. Original building replaced in 1935.	USGS Trout Lake, 6-10E-7, 4 mi. NW of Trout Lake	1922	S	R	190

Department of Natural Resources, Lewis County

No.	Name and Elevation	Map and Location	Built	Removed	Access	Page
273	**Stahl Mt.**—3719 ft.	USGS Morton, 15-4E-35, 10 mi. W of Ashford			AT	190
274	**Pleasant Valley** On Metsker map.	USGS Mineral, 15-4E-36, 9 mi. W of Ashford			R	190
275	**Bald**—3634 ft. On Metsker map.	USGS Morton, 14-3E-5, 12 mi. NW of Morton			R	190
276	**Huckleberry**—3817 ft. Ground house and tower built by Forest Service.	USGS Morton, 14-3E-9, 11 mi. NW of Morton			R	190
277	**Lookout Peak**—3415 ft. May have been used by Huckleberry lookout.	USGS Morton, 14-3E-21, 10 mi. NW of Morton				190

No.	Name and Elevation	Map and Location	Built	Removed	Access	Page
278	**Deschutes (The Rockies)**—4363 ft. In Forest Service records.	USGS Morton, 14-3E-36, 8 mi. NW of Morton		R		190
279	**West Fork**—2000 ft. On Metsker map.	USGS Mineral, 13-4E-12, 4 mi. NW of Morton		R		190
280	**National**—3637 ft. Built by CCC, CRC 50-ft. tower.	USGS Mineral, 14-6E-5, 3 mi. SW of Ashford	1936			190
281	**Ladd Mt.**—4000 ft. Built by SDF, PRC 40-ft. tower, 14×14-ft. building.	USGS Morton, 14-4E-27, 10 mi. W of Ashford	1955	R		190
282	**Kosmos**—3595 ft. Built by SDF, 90-ft. steel tower. Rebuilt on 40-ft. tower in 1962.	USGS Mineral, 12-5E-4, 4 mi. W of Morton	1950	R		190
283	**Kiona**—4819 ft. Built by Forest Service. Moved to Watch Mt.	USGS Randle, 13-6E-35, 3 mi. NW of Randle	1917	R		190
284	**Watch Mt.**—4805 ft. Built by Forest Service. 14×14-ft. ground house relocated in 1963 on 20-ft. tower.	USGS Randle, 13-7E-31, 2 mi. N of Randle	190			
285	**Newaukum Rock**—3700 ft. Old Forest Service photo and on Metsker map.	USGS Morton, 14-3E-31, 10 mi. NW of Morton				190
286	**Coyote Mt.**—2870 ft. Built by SDF, PRC 40-ft. tower, 14×14-ft. building. Spelled Coyotie on Metsker map.	USGS Elk Rock, 11-3E-22, 10 mi. SW of Morton	1955	R		190
287	**Baw Faw (Boistfort Peak)**—3110 ft. Built by CCC, CRC 100-ft. tower, 7×7-ft. building. Building replaced in 1960.	USGS Ryderwood, 12-4W-32, 15 mi. SW of Chehalis	1935	R		187
288	**Lucas Creek**—2116 ft. Shown on Forest Service computer.	USGS Onalaska, 14-1E-35, 13 mi. E of Chehalis		R		190
289	**Devils Burn**—2140 ft. On Metsker map. Does not show on USGS.	USGS Elk Rock, 11-4E-28, 9 mi. S of Morton		R		190
290	**Hunters Cabin**—2608 ft.	USGS Onalaska, 15-1E-36, 14 mi. NE of Chehalis		R		190
291	**Doty**—2081 ft.	USGS Pe Ell, 14-5W-22, 13 mi. W of Chehalis		R		187

Department Of Natural Resources, Cowlitz County

No.	Name and Elevation	Map and Location	Built	Removed	Access	Page
292	**Signal Peak**—3200 ft. 40-ft. tower.	USGS Toutle, 9-2E-3, 15 mi. E of Castle Rock	1959	R		190
293	**Wolf Point**—2996 ft.	USGS Pigeon Springs, 9-2E-32, 15 mi. SE of Castle Rock		R		190
294	**Elk Mt.**—4538 ft. Built by CCC, CRC 28-ft. tower, 14×14-ft. building.	USGS Cougar, 8-3E-32, 9 mi. NW of Cougar	1941	R		190
295	**Davis Peak**—2955 ft. May be another name for Powder Horn Lookout.	USGS Ariel, 6-2E-20, 14 mi. W of Cougar		R		190
296	**Abernathy**—2577 ft. Built by SDF, PRC 40-ft. tower, 14×14-ft. building.	USGS Ryderwood, 10-3W-19, 9 mi. NW of Castle Rock	1951	R		187
297	**Incline**—2291 ft. Built by CCC, NL 40-ft. tower.	USGS Ryderwood, 10-4W-33, 13 mi. NW of Castle Rock	1938	R		187
298	**Gilbert**—4100 ft. RC, 40-ft. tower, 14×14-ft. building. Does not show on USGS.	USGS Elk Rock, 10-4E-31, 26 mi. E of Castle Rock	1963	R		190
299	**Powder Horn**—2954 ft. Built by SDF, smoke house, 10×10-ft. building. Rebuilt in 1960 on 20-ft. tower with 14×14-ft. building. May be another name for Davis Peak.	USGS Ariel, 6-2E-19, 14 mi. NW of Cougar	1946	R		190

Department of Natural Resources, Clark County

No.	Name and Elevation	Map and Location	Built	Removed	Access	Page
300	**Green Knob**—2259 ft. Shown on Forest Service map.	USGS Yacolt, 5-4E-24, 11 mi. S of Cougar		R		190
301	**Larch Mt.**—3496 ft. Built by SDF, PRC 40-ft. tower, 14×14-ft. building.	USGS Camas, 3-4E-27, 18 mi. W of Stevenson	1952			190

Department of Natural Resources, Skamania County

No.	Name and Elevation	Map and Location	Built	Removed	Access	Page
302	**Three Corner Rock**—3300 ft. Built by SDF, 14×14-ft. building rebuilt 1959 and 1972.	USGS Bridal Veil, 3-6E-22, 8 mi. NW of Stevenson	1955	R		190
303	**Greenleaf Peak**—3427 ft.	USGS Bonneville Dam, 3-7E-29, 4 mi. NW of Stevenson		R		190

No.	Name and Elevation	Map and Location	Built	Removed	Access	Page
304	**Lakeview Mt.**—6661 ft. Ground cabin abandoned 1945.	USGS Walupt Lake, 10-11E-2, 24 mi. W of White Swan	1930s	S	RT	190
305	**Jennies Butte**—6410 ft. Wooden tower replaced in 1936 with steel tower.	USGS Jennies Butte, 11-12E-34, 21 mi. W of White Swan	mid-1920s	S	RR	190
306	**Goat Butte**—7407 ft. Ground cabin dynamited by unknown person.	USGS Mt. Adams East, 8-11E-4, 27 mi. SW of White Swan	1930s	c. 1972	RT	190
307	**Signal Peak**—5100 ft. Abandoned.	USGS Signal Peak, 9-13E-36, 17 mi. SW of White Swan	1962 or 1963	S	RR	190
308	**Panther Creek Butte**—4972 ft. Original tower replaced in 1935. Moved to Signal Peak in 1962 or 1963.	USGS Castle Falls, 10-13E-34, 17 mi. SW of White Swan	1917 or 1918	1962 or 1963	RR	190
309	**Simco Butte**—4798 ft. May have been built by DNR and then used by the Yakima Indian Reservation.	USGS Logy Creek, 6-17E-17, 22 mi. S of White Swan		1961 or 1962	RR	190
310	**Satus**—4182 ft. Ground cabin. About 1959 lookout was destroyed in a 100-mi. wind. It was replaced with a steel building, which is still in use.	USGS Fort Simco, 9-16E-24, 6 mi. S of White Swan	early 1930s	S	RR	190
311	**Hagerty Butte**—3602 ft. Steel tower with ground cabin. Used by St. Regis Timber Co. until late 1950s and now abandoned.	USGS Hagerty Butte, 7-15E-31, 19 mi. SW of White Swan	early 1930s	S	RR	190
312	**McKays Butte**—4038 ft. Wooden tower.	USGS McKays Butte, 8-15E-28, 14 mi. SW of White Swan	early 1930s	1961 or 1962	RR	190
313	**Sopelia** 4-ft. tower still in use. Named after a local family.	On Yakima Indian Reservation Exact location unknown		S	RR	190
314	**Sheep Butte**—4422 ft. Platform in tree.	USGS Hagerty Butte, 7-15E-11, 17 mi. SW of White Swan	1961	S	RR	190

Department of Natural Resources, Klickitat County

No.	Name and Elevation	Map and Location	Built	Removed	Access	Page
315	**Shaw Mt. (Meadow Butte)**—3620 ft. Forest Service lists this as Meadow Butte; Metsker map as Shaw Mt.	USGS King Mt., 6-11E-15, 6 mi. E of Trout Lake			R	190
316	**Diamond Cap**—3005 ft. Ground cabin built by CCC.	USGS Husum, 5-12E-15, 23 mi. W of Goldendale	1938		R	190
317	**Grayback Mt.**—3766 ft. Ground cabin built by SDF.	USGS Klickitat, 6-14E-20, 15 mi. NW of Goldendale	1945		R	190
318	**Lorena Butte**—2291 ft. Built by SDF. Now a city park.	USGS Goldendale, 4-16E-27, 1 mi. S of Goldendale	1954		R	190
319	**Nestor** Listed by Forest Service computer but location is in middle of Goldendale, WA.	4-16E-20?				

Wenatchee National Forest, Chelan Ranger District

No.	Name and Elevation	Map and Location	Built	Removed	Access	Page
320	**Stiletto**—7760 ft. 14×14-ft. ground house.	USGS McAlester, 34-17E-1, 13 mi. N of Stehekin	1931	1953	T	188
321	**Goode Ridge**—approx. 6800 ft.	USGS Goode Mt., 34-16E-8, 16 mi. NW of Stehekin	1936	1948	AT	188
322	**McGregor Mt.**—8122 ft. Ground house. Site now has a radio relay station.	USGS McGregor Mt., 34-16E-36, 10 mi. NW of Stehekin	1932	1953	MC	188
323	**Boulder Butte**—7372 ft.	USGS Sun Mt., 33-18E-34, 3 mi. E of Stehekin			AT	188
324	**Horton Butte**—6834 ft.	USGS Lucerne, 32-19E-30, 8 mi. SE of Stehekin	1934	1953	AT	188
325	**Domke Mt.**—4121 ft. On 110-ft. steel tower with ground cabin.	USGS Lucerne, 31-18E-23, 10 mi. SE of Stehekin			AT	188
326	**Vie Mt.**—7300 ft. 10×10-ft. ground house. Not shown on USGS map.	USGS Prince, 31-19E-35, 16 mi. SE of Stehekin	1934	1953	NT	188
327	**Nelson Butte (Knob)**—6267 ft. 10-ft. tower	USGS South Navarre, 30-20E-16, 24 mi. NW of Chelan	1934	1953	R	188
328	**Cooper Mt.**—5867 ft. On 20-ft. tower. Cooper Mountain had first a treehouse and tent, which predated the 1929 Camas fire that covered the mountain. Two firefighters were lost. Building was sold in 1953.	USGS Cooper Mt., 29-22E-16, 12 mi. N of Chelan	1932	1953	R	188

No.	Name and Elevation	Map and Location	Built	Removed	Access	Page
329	**Chelan Butte**—3835 ft. Built by DNR but used by the Forest Service. Moved to Entiat.	USGS Chelan, 27-22E-26, 2 mi. S of Chelan		S	R	189

Wenatchee National Forest, Entiat Ranger District

No.	Name and Elevation	Map and Location	Built	Removed	Access	Page
330	**Duncan Hill**—7819 ft.	USGS Lucerne, 30-18E-31, 18 mi. S of Stehekin	1969		T	188
331	**Pyramid Mt.**—8247 ft.	USGS Lucern, 30-18E-23, 17 mi. S of Stehekin	1932	1954	T	84
332	**Big Hill**—6827 ft.	USGS Big Goat Mt., 29-19E-17, 25 mi. NW of Chelan			R	188
333	**Junior Point**—6654 ft.	USGS Brief, 29-19E-24, 20 mi. NW of Chelan			R	188
334	**Stormy Mt.**—7198 ft. May have been used in 1920 as lookout during emergencies.	USGS Stormy Mt., 28-20E-29, 15 mi. W of Chelan			T	188
335	**Klone Peak**—6834 ft.	USGS Silver Falls, 29-18E-32, 14 mi. N of Plain		c. 1967	T	188
336	**Tyee Mt.**—6654 ft. Ground cabin for emergency use, panorama photo 1934.	USGS Tyee Mt., 27-19E-5, 18 mi. NW of Entiat			R	188
337	**Sugarloaf Peak**—5814 ft. Mention made of use in 1920. Rebuilt 14×14-ft. building in 1944.	USGS Sugarloaf, 26-18E-13, 6 mi. E of Plain		S	R	189
338	**Goman Peak**—3499 ft. Shown on 1937 Forest Service map.	USGS Baldy Mt., 27-21E-30, 9 mi. N of Entiat			R	189
339	**Steliko**—2586 ft. On 10-ft tower, emergency use, panorama photo 1934.	USGS Ardenvoir, 26-20E-20, 8 mi. NW of Entiat		S	R	189
340	**Byrd Point**—3757 ft. Panorama photo 1935.	USGS Baldy Mt., 26-20E-12, 7 mi. NW of Entiat				189
341	**Keystone Point**—3850 ft. Panorama photo 1934.	USGS Ardenvoir, 25-20E-25, 2 mi. SW of Entiat			R	189
342	**Badger Mt.**—3498 ft. This building was moved here from Lion Rock.	USGS Orondo, 23-21E-6, 6 mi. N of Wenatchee	late 1930s	Moved	RR	189

Wenatchee National Forest, Lake Wenatchee Ranger District

No.	Name and Elevation	Map and Location	Built	Removed	Access	Page
343	**Carne Mt.**—7083 ft.	USGS Holden, 30-16E-24, 2 mi. E of Trinity		1955	T	188
344	**Estes Butte**—5942 ft.	USGS Wenatchee Lake, 29-16E-12, 5 mi. SE of Trinity			T	188
345	**Basalt Peak**—6004 ft. Platform, no building, panorama photo 1936.	USGS Chikamin Creek, 29-17E-29, 8 mi. SE of Trinity			T	188
346	**Mt. David**—7420 ft. Stone outhouse still standing	USGS Wenatchee Lake, 29-15E-33, 10 mi. SW of Trinity	1934	after 1972	T	188
347	**Poe Mt.**—6007 ft.	USGS Poe Mt., 28-14E-14, 15 mi. SW of Trinity	early 1930s		T	188
348	**Kodak Peak**—6121 ft. No building, panorama photo 1934.	USGS Bench mark Mt., 29-13E-29, 17 mi. SW of Trinity			T	188
349	**Rock Mt.**—6852 ft. Ground building, panorama photo 1934.	USGS Wenatchee Lake, 27-15E-33, 15 mi. W of Plain		early 1970s	T	189
350	**Alpine**—6237 ft. NHLR Reconstructed 1976, panorama photo 1934.	USGS Wenatchee Lake, 27-16E-28, 10 mi. W of Plain		S	T	189
351	**Dirtyface Peak**—5989 ft. Mention made of use in 1920. Building reconstructed 1956.	USGS Wenatchee Lake, 27-16E-12, 10 mi. NW of Plain		1976	T	188
352	**Cougar Mt.**—6719 ft.	USGS Silver Falls, 28-18E-33, 10 mi. N of Plain			T	188
353	**Soda Springs** No building, panorama photo 1934.	Location unknown				

No.	Name and Elevation	Map and Location	Built	Removed	Access	Page

Wenatchee National Forest, Leavenworth Ranger District

No.	Name and Elevation	Map and Location	Built	Removed	Access	Page
354	**Chumstick Mt.**—5810 ft.	USGS Chumstick Mt., 25-19E-22, 10 mi. NE of Leavenworth	1931	1968	R	189
355	**Burch Mt.**—4522 ft. Site now has a radio relay station.	USGS Rocky Reach Dam, 24-20E-30, 11 mi. SW of Entiat			R	189
356	**Tumwater** Tent in 1920, 1921, and 1922. Also used other years.	USGS Leavenworth, 25-17E, 2-5 mi. N of Leavenworth Exact location unknown				189
357	**Lorraine Point**—5451 ft.	USGS Stevens Pass, 25-13E-13, 21 mi. NW of Leavenworth	1931		T	189
358	**French Ridge**—5800 ft.	USGS Stevens Pass, 25-15E-29, 17 mi. NW of Leavenworth	1934		T	189
359	**McCue Ridge**—5136 ft. On 40-ft. tower.	USGS Chiwaukum Mts., 26-16E-27, 11 mi. NW of Leavenworth	1931		T	189
360	**Grindstone Mt.**—7500 ft. May have only been an observation point.	USGS Chiwaukum Mts., 25-16E-30, 12 mi. NW of Leavenworth			T	189
361	**Icicle Ridge**—7029 ft. Cupola-style building with catwalk.	USGS Chiwaukum Mts., 24-17E-7, 4 mi. NW of Leavenworth	1929	1968	T	189
362	**Boundary Butte**—3168 ft.	USGS Leavenworth, 24-17E-25, 4 mi. S of Leavenworth			R	189
363	**Tiptop**—4766 ft.	USGS Liberty, 23-18E-31, 11 mi. S of Leavenworth	1925	1968	R	189
364	**Three Brothers**—7167 ft.	USGS Liberty, 22-17E-6, 13 mi. SW of Leavenworth	1934	1968	T	189
365	**Beehive**—4576 ft.	USGS Mission Peak, 21-19E-12, 6 mi. SW of Wenatchee	1931		R	189
366	**Mission Peak**—6876 ft.	USGS Mission Peak, 21-19E-27, 11 mi. SW of Wenatchee	1933	probably in 1940s	R	189
367	**Jack Ridge**—6200 ft. Panorama photo 1934.	USGS Jack Ridge, 24-15E-24	1931		T	

Wenatchee National Forest, Cle Elum Ranger District

No.	Name and Elevation	Map and Location	Built	Removed	Access	Page
368	**Fish Lake (Clark Lake)**—3400 ft. On a tower. Panorama photo 1934. Clark Lake was never an official name. Billy Clark was a trapper who lived in the area.	USGS The Cradle, 24-14E-34, 25 mi. NE of Easton		1942	R	189
369	**Davis Peak**—6426 ft.	USGS Kachess Lake, 23-14E-27, 15 mi. NE of Easton	1934	1968	T	189
370	**Polallie Ridge**—approx. 5500 ft. Panorama photo 1934.	USGS Kachess Lake, 23-13E-23, 14 mi. N of Easton	1966		T	189
371	**Red Mt.**—5707 ft.	USGS Kachess Lake, 22-13E-19, 11 mi. N of Easton	mid-1930s	1948	T	189
372	**Thorp Mt.**—5854 ft. National Historic Building.	USGS Kachess Lake, 22-13E-27, 10 mi. N of Easton	1931	S	T	189
373	**Jolly Mt.**—6443 ft. Panorama photo was taken from top of the building in 1928.	USGS Kachess Lake, 22-14E-24, 13 mi. NE of Easton	before 1928	1968 or 1969	T	189
374	**Roaring Ridge**—4988 ft.	USGS Kachess Lake, 22-11E-33, 12 mi. NW of Easton		1967 or 1968	T	189
375	**Martin**—Approx. 2800 ft. Panorama photo 1934.	USGS Snoqualmie Pass, 21-12E-26. 5 mi. W of Easton				189
376	**Goat Peak**—4981 ft. Original ground house replaced by a 12-ft. tower in the late 1950s. Sold and moved.	USGS Easton, 20-13E-27, 4 mi. S of Easton	1920		T	189
377	**Big Creek** On 40-ft tower near North Ridge. May be another name for North Ridge.	USGS Easton			T	
378	**North Ridge**—5268 ft. On 40-ft. tower. Panorama photo 1934.	USGS Easton, 19-13E-12, 7 mi. S of Easton			T	189
379	**Easton Ridge**—2586 ft. On 40-ft. tower. Panorama photo 1935.	USGS Easton, 20-14E-23, 6 mi. SE of Easton				189

No.	Name and Elevation	Map and Location	Built	Removed	Access	Page

Wenatchee National Forest, Ellensburg Ranger District

No.	Name and Elevation	Map and Location	Built	Removed	Access	Page
380	**Koppen Mt.**—6013 ft. May have been an observation point. Panorama photo 1934.	USGS Kachess Lake, 22-15E-9, 14 mi. N of Cle Elum			T	189
381	**Elbow Peak**—5673 ft. May have been an observation point. Panorama photo 1934.	USGS Kachess Lake, 21-15E-6, 16 mi. N of Cle Elum			T	189
382	**Teanaway Butte**—4769 ft. Panorama photo 1936.	USGS Mt. Stuart, 21-15E-10, 9 mi. N of Cle Elum	before 1936	1968	R	189
383	**Stafford**—3784 ft. Panorama photo 1936.	USGS Mt. Stuart, 22-16E-33, 12 mi. NE of Cle Elum		c. 1968	R	189
384	**Standup Ridge** May be another name for Stafford.	USGS Mt. Stuart, 22-16E-?, 12-16 mi. NE of Cle Elum Exact location unknown				
385	**Red Top**—5361 ft. Rebuilt 1962-63, maintained by volunteers.	USGS Mt. Stuart, 21-17E-19, 11 mi. NE of Cle Elum	early 1930s	S	R	189
386	**Lion Rock**—6359 ft.	USGS Liberty, 20-18E-4, 17 mi. N of Ellensburg		1968	R	189
387	**Quartz Mt.**—6290 ft. Panorama photo 1938.	USGS Easton, 18-14E-3, 11 mi. S of Cle Elum		1967	R	189
388	**Frost Mt.**—5740 ft. Emergency use.	USGS Cle Elum, 18-15E-8, 8 mi. S of Cle Elum	c. 1958		T	189
389	**Taneum Point**—3580 ft. 40-ft tower, panorama photo 1934.	USGS Cle Elum, 19-15E-23, 4 mi. S of Cle Elum		1968	R	189

Wenatchee National Forest, Naches Ranger District

No.	Name and Elevation	Map and Location	Built	Removed	Access	Page
390	**Raven Roost**—6198 ft. Site now used for telephone microwave station.	USGS Lester, 18-12E-22, 21 mi. SW of Cle Elum	1934	1964	R	189
391	**Goat Peak (American Ridge)**—6473 ft.	USGS Lester, 17-13E-30, 23 mi. SW of Cle Elum	1934		T	189
392	**Bald Mt.**—5898 ft. Steel tower replaced with wooden tower.	USGS Manastash Lake, 17-15E-15, 15 mi. S of Cle Elum	1933		R	189
393	**Devils Slide**—5625 ft. Tent and firefinder only. Used in conjunction with Bald Mt.	USGS Manastash Lake, 17-15E-17, 15 mi. S of Cle Elum			R	189
394	**Little Bald Mt.**—6108 ft. Accidentally burned in 1980.	USGS Old Scab Mt., 16-13E-2, 25 mi. NW of Naches	1934	1980	R	189
395	**Clover Springs**—6351 ft. Tent and firefinder only. Used in conjunction with Little Bald Mt.	USGS Timber Wolf Mt., 16-13E-22, 25 mi. NW of Naches			R	189
396	**Miners Ridge**—6072 ft.	USGS Bumping Lake, 15-11E-12, 11 mi. N of White Pass	1934		R	189
397	**Mt. Aix**—7772 ft. Probably the first lookout in the district. The trail gains 3800 feet in 7 mi.	USGS Bumping Lake, 15-13E-18, 13 mi. NW of White Pass			T	189

Wenatchee National Forest, Tieton Ranger District

No.	Name and Elevation	Map and Location	Built	Removed	Access	Page
398	**Timberwolf Mt.**—6391 ft.	USGS Timberwolf Mt., 15-13E-25, 20 mi. NW of Naches		1975	R	190
399	**Tumac Mt.**—6340 ft. Built on a cinder cone.	USGS White Pass, 14-12E-8, 5 mi. N of White Pass	1920s	1960s	T	190
400	**Spiral Butte**—5900 ft. Trail to top but no lookout was built.	USGS White Pass, 13-12E-32, 2 mi. NW of White Pass			T	190
401	**Round Mt.**—5971 ft.	USGS White Pass, 13-12E-9, 3 mi. W of White Pass		1976	T	190
402	**Jump Off**—5670 ft. Garage removed 1975. Emergency use.	USGS Tieton Basin, 13-14E-1, 15 mi. SW of Naches		S	R	190
403	**Bear Creek Mt.**—7336 ft.	USGS White Pass, 12-12E-17, 9 mi. S of White Pass			AR	190
404	**Darland**—6981 ft.	USGS Darland Mt., 12-13E-20, 12 mi. SE of White Pass			R	190
405	**Blue Slide**—6785 ft.	USGS Darland Mt., 12-13E-4, 11 mi. SE of White Pass			R	190

No.	Name and Elevation	Map and Location	Built	Removed	Access	Page

Department of Natural Resources, Chelan County

No.	Name and Elevation	Map and Location	Built	Removed	Access	Page
406	**Naneum Point**—6623 ft. PRC 40-ft. tower, 14×14-ft. building.	USGS Mission Peak, 21-19E-36, 11 mi. SW of Wenatchee	1958		R	189
407	**Harriets Peak**—5459 ft. Manned by Forest Service. Doesn't show on USGS.	USGS Liberty Peak, 21-19E-19, 11 mi. SW of Wenatchee			R	189

Department of Natural Resources, Kittitas County

No.	Name and Elevation	Map and Location	Built	Removed	Access	Page
408	**Peoh Point**—4020 ft. Built by SDF. Ground house rebuilt about 1978.	USGS Cle Elum, 19-15E-10, 2 mi. S of Cle Elum	1951	1985	R	189
409	**Manastash**—3930 ft. Built by SDF, 10×10-ft. smoke house.	USGS Ellensburg, 17-17E-21, 9 mi. SW of Ellensburg	1942		R	189
410	**Middle Teanaway** Panorama photo shows lookout construction.	Location unknown			R	

Department of Natural Resources, Yakima County

No.	Name and Elevation	Map and Location	Built	Removed	Access	Page
411	**Cleman Mt.**—5115 ft. Built by SDF. Ground house was rebuilt on 40-ft. tower in 1968.	USGS Milk Canyon, 16-16E-32, 17 mi. NW of Yakima	1954		R	189
412	**Pine Mt.**—4300 ft. 14×14-ft. building on 40-ft. tower.	USGS Pine Mt., 13-15E-25, 13 mi. W of Yakima	1961		R	190
413	**Sedge**—4837 ft. On Sedge Ridge.	USGS Pine Mt., 12-15E-19, 16 mi. SW of Yakima			R	190

Okanogan National Forest, Conconully Ranger District

No.	Name and Elevation	Map and Location	Built	Removed	Access	Page
414	**Windy Peak**—8334 ft. 14×14-ft. ground house.	USGS Horseshoe Basin, 40-23E-33, 16 mi. NW of Loomis	1932		T	188
415	**Juniper Point**—5200 ft. Platform only, still standing.	USGS Horseshoe Basin, 39-24E-13, 7 mi. NW of Loomis			R	188
416	**Corral Butte**—6849 ft. Ground house.	USGS Horseshoe Basin, 38-23E-3, 14 mi. W of Loomis	1933	1957	R	188
417	**Thunder Mt.**—7083 ft.	USGS Horseshoe Basin, 38-23E-26, 14 mi. SW of Loomis			AT	188
418	**Mt. Tiffany**—8242 ft.	USGS Tiffany Mt., 37-23E-27, 12 mi. NW of Conconully	1932	1953		188
419	**Old Baldy**—7844 ft. Ground house built by CCC.	USGS Tiffany Mt., 36-23E-28, 10 mi. W of Conconully	1934		AT	188
420	**Muckamuck**—6370 ft. Old cupola building still in fair condition in 1968 and platform may still be standing.	USGS Tiffany Mt., 36-24E-20, 6 mi. NW of Conconully	1932		AT	188
421	**Funk Mt.**—5122 ft.	USGS Conconully, 36-25E-19, 8 mi. N of Conconully	1932	S	R	188
422	**Starvation Mt.**—6769 ft. Ranger said there is evidence that a building was here, but he does not know its purpose.	USGS Tiffany Mt., 35-23E-15, 9 mi. SW of Conconully			R	188
423	**Granite Mt.**—7366 ft. Ground house.	USGS Loup Loup, 35-23E-25, 8 mi. SW of Conconully	1932	1954	AT	188
424	**Jackass Butte** Stood on a 6-ft. foundation.	Location unknown	1935	1957		

Okanogan National Forest, Winthrop Ranger District

No.	Name and Elevation	Map and Location	Built	Removed	Access	Page
425	**Monument 83**—approx. 6550 ft. Original log cabin still standing. Lookout on 30-ft. tower was built in 1953.	USGS Frosty Creek, 40-18E-5, 33 mi. NW of Mazama	1930	S	T	188
426	**Bunker Hill**—7239 ft. Last used in 1966.	USGS Ashnola Mt., 40-19E-15, 27 mi. N of Mazama	1932		T	188
427	**Point Defiance**—7403 ft.	USGS Mt. Lago, 39-18E-15, 22 mi. NW of Mazama	1932	1954	T	188
428	**Dollar Watch**—7694 ft. Tent in the 1920s.	USGS Ashnola Mt., 39-20E-7, 21 mi. N of Mazama	1932		T	188
429	**Diamond Point**—7916 ft. 14×14-ft. ground house.	USGS Ashnola Pass, 39-20E-14, 20 mi. NE of Mazama	1932	1952	T	188
430	**Remmel Mt.**—8685 ft. 14×14-ft. ground house.	USGS Remmel Mt., 40-21E-34, 26 mi. NW of Loomis	1932	1952	T	188

No.	Name and Elevation	Map and Location	Built	Removed	Access	Page
431	**Crater Mt.**—8128 ft.	USGS Crater Mt., 37-16E-5, 16 mi. NE of Newhalem	1932	1968	MC	188
432	**North Crater Mt.**—7054 ft. Never used. Baker River Ranger Station's records show it was built in 1938, but central Forest Service records show 1955.	USGS Crater Mt., 37-16E-5, 17 mi. NE of Newhalem	1955	1972	T	188
433	**Cady Point**—6542 ft. 10×10-ft. building.	USGS Azurite Peak, 37-17E-6, 18 mi. W of Mazama	mid-1930s		AT	188
434	**Slate Peak**—7440 ft. Lookout, on 50-ft. tower, is used during emergencies. Peak was originally 7488 ft. high, but top 48 ft. were removed for an Army building.	USGS State Peak, 37-17E-1, 17 mi. NW of Mazama		S	RR	188
435	**Mebee Pass**—approx. 6500 ft.	USGS Azurite Peak, 36-17E-18, 18 mi. NW of Mazama	1934	1954	AT	188
436	**Driveway Butte**—5982 ft. 30-ft. tower. A tree platform was used before the building.	USGS Robinson Mt., 36-19E-7, 6 mi. NW of Mazama	1934	1953	T	188
437	**Mt. Setting Sun**—7253 ft. Built by CCC.	USGS Mazama, 37-19E-23, 7 mi. N of Mazama	1934	1953	AT	188
438	**Goat Peak**—7001 ft. Reconstructed in 1948 on a 10-ft. tower. Emergency use.	USGS Mazama, 36-19E-12, 2 mi. N of Mazama	1932	S	T	188
439	**Milton Mt.**—7152 ft. Sold and removed in 1951.	USGS Buttermilk Butte, 35-19E-35, 12 mi. W of Winthrop	1933	1951	AT	188
440	**Sweetgrass Butte**—6104 ft. Last manned 1964. Part of AWS.	USGS Mazama, 37-20E-35, 15 mi. NW of Winthrop	1933	1970	R	188
441	**Burch Mt.**—7782 ft. Built by CCC on a 28-ft. tower.	USGS Billy Goat Mt., 38-20E-14, 22 mi. N of Winthrop	1934	c. 1960	T	188
442	**North Twentymile Peak**—7437 ft. 40-ft tower. Older 12×12-ft. building with cupola standing nearby. Still used.	USGS Doe Mt., 38-22E-34, 20 mi. N of Winthrop	1923	S	T	188
443	**Doe Mt.**—7154 ft.	USGS Does Mt., 37-21E-11, 17 mi. N of Winthrop	1935	1952		188
444	**First Butte**—5491 ft. 35-ft. tower. Still used.	USGS Doe Mt., 36-22E-17, 10 mi. N of Winthrop	1937	S	R	188
445	**Pearrygin Peak**—6644 ft. 40-ft. tower.	USGS Doe Mt., 35-22E-13, 8 mi. NE of Winthrop	1935	1953	AT	188
446	**Winthrop Butte (Stud Horse Butte)** Firefinder on stump used in 1950s.	USGS Winthrop, 34-21E-1, 1 mi. NE of Winthrop Exact location uncertain			R	188
447	**Gardner Mt.**—8877 ft. Shown on DNR records but nothing known.	USGS Mazama, 35-19E-28, 14 mi. W of Winthrop			NT	188

Okanogan National Forest, Twisp Ranger District

No.	Name and Elevation	Map and Location	Built	Removed	Access	Page
448	**North Creek** 1934 panorama photo shows top of building.	USGS Gilbert, 34-19E-?, 20 mi. NW of Twisp Exact location unknown			NT	188
449	**South Creek Butte**—7670 ft. 10×10-ft. ground house.	USGS Gilbert, 34-18E-21, 23 mi. W of Twisp	1935		NT	188
450	**Midnight Mt.**—7480 ft. 1934 panorama photo shows elaborate rock wall lining trail to lookout.	USGS Buttermilk Butte, 34-19E-23, 17 mi. W of Twisp	1932	1952	T	188
451	**War Creek Ridge** Ground house, sold and removed.	USGS Buttermilk Butte, 33-19E-?, 17 mi. W of Twisp Exact location unknown	1940	1951	AT	188
452	**Buttermilk Butte**—5471 ft. Ground house, sold and removed.	USGS Buttermilk Butte, 32-20E-2, 10 mi. SW of Twisp	1933	1951	R	188
453	**Lookout Mt.**—5515 ft. Built 30-ft. tower in 1937. Still used.	USGS Twisp West, 33-21E-35, 5 mi. SW of Twisp	1913 or 1914	S	T	188
454	**Gold Ridge** Tree tower used by Foggy Dew Guard Station.	USGS Hungry Mt., 31-21E-?, 10 mi. S of Twisp Exact location unknown				188
455	**Mt. Leecher**—5012 ft. There are three generations of lookout buildings on this spot.	USGS Twisp East, 32-23E-30, 9 mi. SE of Twisp	1921		R	188

456 **Thrapp Mt.**—4266 ft. 1934 panorama photo.	USGS Loup Loup, 32-23E-1, 11 mi. SE of Twisp	1933	1957	R	188
457 **Little Buck Mt.**—5390 ft. Built by CCC. 1934 panorama photo.	USGS Loup Loup, 33-23E-2, 9 mi. E of Twisp	1934	1953	R	188
458 **Boulder** 1934 photo of lookout exists.	Location unknown				
459 **Black Ridge** 1934 panorama photo.	USGS Buttermilk Butte, 33-19E-?, 17 mi. W of Winthrop Location unknown	1935	1952		188
460 **McClure Mt.**—4600 ft. Shown on DNR records.	USGS Twisp East, 33-22E-32, 3 mi. S of Twisp			NT	188
461 **Oval Peak**—8795 ft. Shown on DNR records.	USGS Oval Peak, 32-19E-11, 14 mi. SW of Twisp			NT	188

Department of Natural Resources, Okanogan County

462 **Lemanasky (Aeneas Mt.)**—5167 ft. Built by Forest Service but transferred to DNR in 1958. PRC 40-ft. tower, 14×14-ft. building. New building, 1980.	USGS Conconully, 38-26E-31, 5 mi. S of Loomis	1934 1980	S	RR	188
463 **Skull and Crossbones**—6717 ft. Built by Forest Service but sold to DNR in 1954.	USGS Horseshoe Basin, 38-24E-28, 9 mi. SW of Loomis	1944		R	188
464 **Buck Mt.**—6135 ft. Built by Forest Service. 20-ft. pole, rebuilt 1960 on 40-ft. tower, 14×14-ft. building.	USGS Loup Loup, 34-24E-21, 12 mi. W of Okanogan	1935	S	RR	188
465 **Chilwist Butte**—3071 ft.	USGS Okanogan, 32-25E-6, 8 mi. SW of Okanogan			R	188
466 **Knowlton Knob**—3852 ft. Built by Forest Service. Ground house sold to state in 1954. Rebuilt on 40-ft. tower in 1960. Staffed by Gebbers Timber Co.	USGS Brewster, 31-23E-24, 7 mi. NW of Brewster	1940		R	188

DIRECTIONS TO SELECTED LOOKOUTS

The following are general directions to some of the more popular and accessible lookouts described in this book. Access can change so you should write or call the appropriate park or ranger district for updated information. It is always a good idea to call ahead for information on road and weather conditions. Detailed descriptions of hikes to these sites can be found in *50 Hikes in*™ *Mount Rainier National Park*, *100 Hikes in*™ *Washington's North Cascades*, *100 Hikes in*™ *Washington's Alpine Lakes*, and *100 Hikes in*™ *Washington's Glacier Peak Region*, all published by The Mountaineers.

Mount Baker–Snoqualmie National Forest

BAKER RANGER DISTRICT
2105 Highway 20
Sedro Woolley, WA 98284
Phone: (360) 856-5700

Church Mountain. 4-mile hike. Drive Mount Baker Highway 542 to 5.1 miles beyond Glacier Public Service Center, then go left on Road 3040 another 2.6 miles to the trailhead.

Winchester Mountain. 2-mile hike. Drive Mount Baker Highway 542 some 12.8 miles beyond Glacier Public Service Center, then go left on Road 3065. Depending on road conditions drive 7 miles to Twin Lakes.

Park Butte. 3½-mile hike. Drive North Cascades Highway 20 east from Sedro Woolley for 14.5 miles, go left for 12.5 miles on the Baker Lake Road, left for 3 miles on Road 12, and then take Road 13 to the trailhead.

Sauk Mountain. Steep 2-mile hike. The lookout is gone but the views and flowers are spectacular. Drive North Cascades Highway 20 east of Concrete for about 10 miles, go left on Road 1030 (next to Rockport State Park) for 7 miles, and then right on Road 1036 to the trailhead.

Lookout Mountain. 4¼-mile hike gaining 4,500 feet. Few make the strenuous climb to the lookout, but those who do rave about the view. Drive North Cascades Highway 20 to Marblemount, then cross the river on the Cascade River Road and drive 7.3 miles to the trailhead.

Hidden Lake Peaks. 4-mile hike. Drive North Cascades Highway 20 to Marblemount, then cross the river on the Cascade River Road, go 10 miles, then go left for 4.7 miles on Road 1540 to the trailhead.

Copper Mountain. 10-mile hike. Drive Mount Baker Highway 542 some 12.8 miles beyond Glacier Public Service Center, then go left for 4 miles on Road 32 to the trailhead.

Sourdough Mountain. 5½-mile hike. Drive North Cascades Highway 20 to Seattle City Light's town of Diablo. Find the trailhead near the swimming pool.

Desolation Peak. 4½-mile hike. Drive North Cascades Highway 20 to Ross Lake. Either take a water taxi to the trailhead or hike the East Bank Trail for 18 miles to the trailhead.

DARRINGTON RANGER DISTRICT
1405 Emmens Street
Darrington, WA 98241
Phone: (360) 436-1155

North Mountain. Road access. Drive Highway 530 to Darrington. A long mile beyond the town's center turn left on Road 28 for about 3 miles, then right on Road 2810 for another 9 miles to the lookout tower.

Green Mountain. 4-mile hike. Between Darrington and Rockport on Highway 530 near the Sauk River bridge, go north on Road 26 for 19 miles, then left on Road 2680 for another 5 miles to the trailhead.

Miners Ridge. 16-mile hike. On Highway 530, between Darrington and Rockport, near the Sauk River Bridge, go north on Road 26 for 23 miles to the Suiattle River trailhead. Hike the river trail for 9.5 miles and then turn left onto Trails 785 and 785A.

Mount Pilchuck. 2-mile hike. Drive the Mountain Loop Highway through Granite Falls, and 1 mile beyond the Verlot Public Service Center, go right on Road 42 for 6.2 miles to the trailhead.

Three Fingers. *For experienced hikers only.* Drive the Mountain Loop Highway for 6.5 miles east of Granite Falls and go left on Road 41 for some 18 miles to the trailhead. At 4.5 miles reach Goat Flats. At about 8 miles reach Tin Can Gap and trail's end. From here are snowfields and rock scrambles.

SKYKOMISH RANGER DISTRICT
P.O. Box 305
Skykomish, WA 98288
Phone: (360) 677-2414

Evergreen Mountain. 1½-mile hike. Drive US Highway 2 to Skykomish. A short mile east of the town go left on Road 65 for some 12 miles and then right on Road 6564. If possible (it probably won't be) drive another 9 miles to the trailhead.

NORTH BEND RANGER DISTRICT
42404 S.E. North Bend Way
North Bend, WA 98045
Phone: (360) 888-1421

Granite Mountain. 4-mile hike. Drive I-90 to Exit 47. On the north side of the freeway turn west for 0.5 mile to the trailhead. Start on Trail 1007 then go right on Trail 1016.

WHITE RIVER RANGER DISTRICT
857 Roosevelt Avenue E.
Enumclaw, WA 98022
Phone: (360) 825-2571

Sun Top. Road access. Drive US Highway 410 east from Enumclaw for 6+ miles past Greenwater and go left on Road 73. In a short 2 miles go right on Road 7315 to the lookout.

MOUNT RAINIER NATIONAL PARK
You can call the park at (360) 569-2211 for more information about the lookouts as well as for updates on weather and road conditions.

Gobblers Knob. 2½-mile hike. Drive to the Nisqually River entrance. One mile inside the park, go left on the West Side Road for another mile to where repeated floods have destroyed the road. Walk or bicycle 6 miles to the trailhead.

Tolmie Peak. 3½-mile hike. Drive to Mowich Lake by way of Road 162 from Buckley and then Road 165 through Wilkeson. Find the trailhead at the lake.

Mount Fremont. 2½-mile hike. Drive Road 410 to the White River entrance and on to the trailhead at the Sunrise Visitor Center.

OLYMPIC NATIONAL FOREST
Sol Duc Ranger District
Rural Route 1, Box 5750, Hwy 101
Forks, WA 98331
Phone: (360) 374-6522

Kloshe Nanich. Road access. Drive US Highway 101 between Lake Crescent and Forks. Near the Sol Duc River bridge take the Snider Work Center road and follow Road 3040 for some 5 miles to the lookout.

OLYMPIC NATIONAL PARK
600 E. Park Avenue
Port Angeles, WA 98362
Phone: (360) 452-4501

Hurricane Hill. 1¼-mile hike. From Port Angeles drive the Hurricane Ridge Road. Pass the visitor center to the road-end trailhead.

GIFFORD PINCHOT NATIONAL FOREST
Randle Ranger District
10024 US Highway 12
Randle, WA 98377
Phone: (360) 497-1100

Burley Mountain. ½-mile hike. From Randle drive south on Road 25 to Iron Creek Campground and go straight ahead on Road 76 for 4.5 miles, right on Road 77 for 7.6 miles, then left on Road 7605 for another 1.5 miles, and finally right on Road 7605-086 to the trailhead.

PACKWOOD RANGER DISTRICT
13068 US Highway 12
Packwood, WA 98361
Phone: (360) 494-5515

High Rock. 1¾-mile hike. Drive toward the Nisqually entrance to Mount Rainier. At 3.4 miles beyond Ashford go right on Kernahan Road 52 signed "Packwood." Cross the Nisqually River and in 1.5 miles go straight ahead on Road 85 for about 6 miles, then drive 5 more miles on Road 8440 to the trailhead.

MOUNT ADAMS RANGER DISTRICT
2455 Highway 141
Trout Lake, WA 98650
Phone: (509) 395-2501

Red Mountain. Road access. From the Columbia River drive to Carson. Drive north for 10 miles on the Wind River Highway. Go right for 12 miles on Road 60 to Red Mountain Road 6048. The road is very steep and rough and at times gated. Be prepared to walk.

Mount Adams. *This is a mountain climb.* The mountain is reached from Trout Lake by way of Road 80. One must be experienced in the use of ice axes and crampons and prepared for sudden storms.

Wenatchee National Forest

LEAVENWORTH RANGER DISTRICT
600 Sherbourne
Leavenworth, WA 98826
Phone: (509) 782-1413

Sugarloaf Peak. Road access. From Leavenworth drive US Highway 209 north for some 2 miles, then right for about 7 miles on Eagle Creek Road 7520 to the ridge top, then north for 4 long miles on Road 5200 to the lookout.

Thorp Mountain. Drive the Salmon la Sac Road to French Cabin Creek Road 4308, turn left on it for 5 miles, then turn right onto Knox Creek Road 4308-120 for 2 miles to the trailhead.

Okanogan National Forest

METHOW RANGER DISTRICT
P.O. Box 579
24 W. Chewuch
Winthrop, WA 98862
Phone: (509) 996-2266

Monument 83. 10-mile hike. The lookout is best reached from Canada's Manning Provincial Park on Highway 3. Find the trailhead 1.8 miles east of the park headquarters.

Slate Peak. 1-mile road walk. From North Cascades Highway 20 near Winthrop drive to Mazama and then take Road 54 for 20 miles to Harts Pass. Go left on Slate Peak Road 5400-600 for 1.7 miles to the gate.

North Twentymile Peak. 7½-mile hike. From Winthrop take the Chewack River Road for 7 miles to a junction on the north side of the river. Follow the river road another 11.3 miles and turn right on Road 5010. After 2 miles turn onto Road 5010-560, which takes you to the trailhead.

Lookout Mountain. 1½-mile hike. From Twisp drive the Twisp River Road for about 0.2 mile, then turn left on Road 200 for some 8 miles to the trailhead.

TONASKET RANGER DISTRICT
P.O. Box 466
1 W. Winsap
Tonasket, WA 98855
Phone: (509) 486-2186

Mount Bonaparte. 4½-mile hike. From Tonasket drive east on Road 9467 for 15.7 miles. Turn right on Lost Lake Road and drive 0.9 mile, then turn right again onto Road 3300-300 for 4.2 miles. From there, it is another 1.2 miles to the trailhead.

PRESERVING WESTERN WASHINGTON LOOKOUTS

The following are names and addresses of organizations active in preserving and maintaining historic lookouts. They are involved in restoration of buildings and trail maintenance or devoted to preserving the history of the sites through educational programs. All need and welcome new volunteers.

Forest Fire Lookout Association (FFLA)

Founded in January 1990 at Pennsylvania's Hopewell Forest Fire Station and Tower with Stephen Cummings as president, the organization now has a 600+ membership.

The FFLA is dedicated to the preservation of fire lookouts worldwide. It aims to increase public awareness of these historic sites, provide an opportunity for present and former firewatchers to share experiences, and protect and restore old lookouts. The association has thirty directors in twenty-seven states, plus two in Canada and one in Australia.

The FFLA has a Washington State chapter, headed by Ray Kresek, author of *Fire Lookouts of the Northwest* and *Fire Lookouts of Oregon and Washington*. Kresek spent thirty years compiling these books.

In 1969 Kresek also created a private Fire Lookout Museum in his backyard. Over the years his small museum has grown to cover an entire acre and has become one of the country's most comprehensive collections of lookout history. Admission is free, by appointment.

Both the National FFLA and the Washington chapter hold meetings, which are well attended. An eastern regional and western regional conference are held each year. For more information, contact:

FFLA/Washington State Chapter
c/o W. 123rd Westview
Spokane, WA 99218
(509) 466-9171

FFLA/Western Washington Chapter
c/o P.O. Box 43
Snohomish, WA 98291
(206) 487-3461

217 Preserving Western Washington Lookouts

Columbia Breaks Fire Interpretive Center (CBFIC)

In recent years some of the worst Northwest forest fires have occurred in the Wenatchee National Forest.

Starting first with an inspiration to move an abandoned lookout tower from the mountain to where tourists and schoolchildren could climb the stairs and look for smoke through the firefinder, the inspiration grew and the non-profit Columbia Breaks Fire Interpretive Center Foundation was started in 1990 by Nancy Bell of Wenatchee.

The CBFIC is turning into a full-fledged interpretive center with multimedia displays. The organization trucked down Badger Mountain Lookout, from across the river, and the Chelan Butte Lookout. It is the only interpretive center in the country devoted to wildfires. For more information contact:

> Columbia Breaks Fire Interpretive Center
> P.O. Box 3773
> Wenatchee, WA 98807
> (509) 784-1203

The Skagit Valley Alpine Club

The club's Cabin Committee has been active in maintaining Park Butte Lookout for the past 35 years. Although strictly a volunteer organization, they work closely with the Forest Service, in particular with the ranger district of the Mount Baker–Snoqualmie National Forest. The club does regular cleaning and maintenance of the structure. A Forest Service grant recently enabled a more complex and expensive project (foundation replacement) to be done. For more information, contact:

> Dr. Fred Darvill, Lookout Cabin Committee Chairman
> Skagit Alpine Club
> P.O. Box 513
> Mount Vernon, WA 98273
> (360) 242-5854

Friends of the Hidden Lake Lookout

This organization is a committee of the Skagit Environmental Council, Incorporated. Volunteers assist primarily with cleaning, supplies, and maintenance of the 64-year-old building, and also do some trail work from time to time. New members are welcome. Porters and carpenters are always needed. For more information, contact:

> Dr. Fred Darvill, Chair
> Friends of the Hidden Lake Lookout
> Skagit Environmental Council Incorporated
> 1819 Hickox Road
> Mount Vernon, WA 98273

The Mountaineers, Everett Branch

The Everett, Washington branch of The Mountaineers has an active Lookout/Trail Maintenance Committee. The group supports nine or ten trail preservation projects each year, and the annual "National Trails Day" trail maintenance project attracts more than 125 volunteers.

Pilchuck Lookout was restored through the committee's efforts, and the organization continues to do maintenance work on the structure. More than 100 volunteers put in over 10,000 hours on the project. The group also assisted King County Search and Rescue and the Boy Scouts Search and Rescue Explorer Post with the restoration of Evergreen Mountain Lookout. The committee has also been active in upkeeping Three Fingers Lookout. Their latest project is the restoration of Heybrook Lookout, for which they continue to seek financial assistance and volunteer labor. For more information, contact:

> Lookout/Trail Maintenance Committee
> The Everett Mountaineers
> P.O. Box 1848
> Everett, WA 98026
> (206) 487-3461

INDEX

ABOUT THE AUTHORS

IRA SPRING is known nationwide as half of an outdoor photography team—Bob and Ira Spring—whose work has appeared in numerous national magazines, been displayed in murals at Grand Central Station in New York, and published in more than 50 books. The twin brothers like to say they got their start the year that Eastman Kodak presented a Brownie box camera to every 12-year-old in America and, unlike most of the others, the Springs never tired of taking pictures. While Bob has come to specialize in travel subjects, Ira delights in capturing outdoor and wildlife scenes wherever they occur.

Ira is the author of a series of best-selling hiking guides published by The Mountaineers. When he's not off taking pictures in places as far apart as the Pacific Northwest and the Alps, Ira hangs up his backpack in Edmonds, Washington.

BYRON FISH began backpacking over mountain trails as a teenageer in the 1920s. Ten years later he was hiking in order to write magazine articles about U.S. Forest Service activities. He also took a turn on the payroll as a smokechaser, but paid or not, he wrote about the trails, lookouts, and ranger station activities. One of his articles about lookouts appeared in *The Saturday Evening Post* in 1937.

The "By [his mark] Fish" signature appeared for years on a column in the *Seattle Times*, and regularly in *Family Circle* and other general interest publications. He authored eleven books on such diverse subjects as mountain trails, the scenic attributes of Washington State, and the life of an elephant trainer. Byron Fish passed away in 1996.

By ⟶ HIS MARK

Other titles you may enjoy from The Mountaineers:

IMPRESSIONS OF THE NORTH CASCADES: Essays about a Northwest Landscape, *John C. Miles, editor*
A diverse collection of original essays explores Washington's North Cascades to create a unique portrait of a changing landscape.

MONTE CRISTO, *Philip R. Woodhouse*
The complete story of the Monte Cristo region of the Cascades during the search for gold and silver in its fabled mines in the late 1800s and early 1900s.

SNOQUALMIE PASS: From Indian Trail to Interstate, *Yvonne Prater*
Colorful history of the Washington Cascades pass. Published in conjunction with the Mountains to Sound Greenway Trust.

STEVENS PASS: The Story of Railroading and Recreation in the North Cascades, *JoAnn Roe*
Covers the exploration and development of rails and roads to scenic and recreational areas in this region of Washington State.

EXPLORING WASHINGTON'S WILD AREAS: A Guide for Hikers, Backpackers, Climbers, X-C Skiers, & Paddlers, *Marge & Ted Mueller*
Guide to 55 wilderness areas with outstanding recreational opportunities, plus notes on history, geology, and wildlife.

HIKING THE MOUNTAINS TO SOUND GREENWAY,
Harvey Manning
Recreational walks and all-day hikes along Puget Sound's I-90 corridor. Includes the history, founding, and future of the Greenway project.

OLYMPIC MOUNTAINS TRAIL GUIDE, 2nd Ed., *Robert L. Wood*
Revised guide to every trail in the Olympics, including scenic and historic highlights, mileages, and elevations.

THE IRON GOAT TRAIL, *Volunteers for Outdoor Washington, USDA Forest Service, & Mount Baker-Snoqualmie National Forest*
History-filled walking guide to the first railroad route across the Cascades.

100 HIKES IN WASHINGTON'S GLACIER PEAK REGION: THE NORTH CASCADES, 2nd Ed., *Ira Spring & Harvey Manning*

100 HIKES IN WASHINGTON'S NORTH CASCADES NATIONAL PARK REGION, 2nd Ed., *Ira Spring & Harvey Manning*

100 HIKES IN WASHINGTON'S SOUTH CASCADES AND OLYMPICS, 2nd Ed., *Ira Spring & Harvey Manning*

100 HIKES IN WASHINGTON'S ALPINE LAKES, 2nd Ed., *Ira Spring & Harvey Manning*

THE MOUNTAINEERS, founded in 1906, is a nonprofit outdoor activity and conservation club, whose mission is "to explore, study, preserve, and enjoy the natural beauty of the outdoors. . . ." Based in Seattle, Washington, the club is now the third-largest such organization in the United States, with 15,000 members and five branches throughout Washington State.

The Mountaineers sponsors both classes and year-round outdoor activities in the Pacific Northwest, which include hiking, mountain climbing, ski-touring, snowshoeing, bicycling, camping, kayaking and canoeing, nature study, sailing, and adventure travel. The club's conservation division supports environmental causes through educational activities, sponsoring legislation, and presenting informational programs. All club activities are led by skilled, experienced volunteers, who are dedicated to promoting safe and responsible enjoyment and preservation of the outdoors.

If you would like to participate in these organized outdoor activities or the club's programs, consider a membership in The Mountaineers. For information and an application, write or call The Mountaineers, Club Headquarters, 300 Third Avenue West, Seattle, Washington 98119; (206) 284-6310; e-mail: clubmail@mountaineers.org.

The Mountaineers Books, an active, nonprofit publishing program of the club, produces guidebooks, instructional texts, historical works, natural history guides, and works on environmental conservation. All books produced by The Mountaineers are aimed at fulfilling the club's mission.

Send or call for our catalog of more than 300 outdoor titles:

The Mountaineers Books
1001 SW Klickitat Way, Suite 201
Seattle, WA 98134
1-800-553-4453 / e-mail: mbooks@mountaineers.org